Hydrogenation and hydrogenolysis in synthetic organic chemistry

W0079160

A.P.G. Kieboom
F. van Rantwijk

Delft University Press

Hydrogenation and hydrogenolysis in synthetic organic chemistry

A.P.G. Kieboom
F. van Rantwijk

with a foreword by
H. van Bekkum

1977
DELFT UNIVERSITY PRESS

A.P.G. Kieboom
F. van Rantwijk
Laboratory of Organic Chemistry
Delft University of Technology
Julianalaan 136
Delft-2208
The Netherlands

Copyright © 1977 by Nijgh-Wolters-Noordhoff Universitaire Uitgevers B.V., Rotterdam.
No part of this book may be reproduced in any form by print, photoprint, microfilm or
any other means without written permission from the publisher.
Softcover reprint of the hardcover 1st edition 1977

ISBN-13: 978-90-298-0101-0   e-ISBN-13: 978-94-011-7589-0
DOI: 10.1007/978-94-011-7589-0

Hydrogenation and hydrogenolysis in synthetic organic chemistry

Contents

Foreword

Numerous examples are known of the application of catalyzed hydrogenation and hydrogenolysis reactions in synthetic organic chemistry.

However, catalyst and reaction conditions are often chosen on the analogy of literature data without a good knowledge of the influence of reaction variables and of the structure of the reactant on the various possible modes of reaction. In order to improve such an intuitive procedure it is essential to dispose of an understanding of the reaction mechanisms which are operative in hydrogenation and hydrogenolysis reactions and which govern *i.a.* the selectivity of a given pathway with respect to consecutive and parallel reactions.

Although organic chemistry and, in particular, hydrogenation and hydrogenolysis reactions remain an experimental science this book has been written to give the organic chemist the insight and know-how necessary to apply these reactions successfully to synthetic problems. I warmly recommend it as a book which will enable the organic chemist to rationalize many of the phenomena of catalytic hydrogenation and hydrogenolysis reactions, whereby it surely helps the organic chemist to solve forthcoming synthetic problems in this field. In addition, it seems to me a useful book for physical and physical-organic chemists working in the field of homogeneous and heterogeneous catalysis.

Finally, the ten years' experience of both authors in the field of catalysis and synthetic organic chemistry – as reflected by some forty publications – guarantees a well-considered review containing many examples directly from the bench.

Delft, August 1976.   H. VAN BEKKUM

Preface

The major aim of this book is to provide preparative organic chemists with the insight and the know-how necessary to apply catalytic hydrogenation and hydrogenolysis to synthetic problems.

Several texts on hydrogenation and hydrogenolysis are available, but the authors feel that many chemists will welcome a book in which more attention has been paid to the mechanistic background of these reactions and its relation to synthetic problems. In this book a special effort has been made to present the various types of hydrogenation and hydrogenolysis reactions from both a mechanistic and a preparative point of view.

After a short general introduction concerning catalyst systems and reaction conditions, hydrogenation and hydrogenolysis are discussed separately. The chapters have been organized according to a logical arrangement of the various bonds which can be reduced with hydrogen in the presence of a catalyst system. Reaction rate, selectivity, and steric course of the hydrogen addition are dealt with in relation to the reaction mechanism. Numerous synthetically interesting examples exemplify these aspects as well as the scope and limitations of the reactions.

We wish to thank Mr. J.M. Dijksman for preparing the drawings, and Mrs. T.M. van Linge-Scholten for typing the manuscript. Permission of the Chemical Society and Dr. J. Chatt to reproduce the figure at page 5 is acknowledged. Finally, we are grateful to professor H. van Bekkum for helpful discussions and to Dr. J.M.A. Baas, Mr. D.A. Hoogwater, and Mr. J.A. Peters for reading the proof.

Delft, August 1976.

A.P.G. KIEBOOM
F. VAN RANTWIJK

I. Introduction

1. The reactions

The reactions which form the subject of this book are the catalytic addition of hydrogen to π-bonds (hydrogenation, 1) and the catalytic reductive cleavage of σ-bonds (hydrogenolysis, 2).

$$A=B \ + \ H-H \ \longrightarrow \ HA-BH \qquad (1)$$

$$A,B \ = \ C,N,O$$

$$A-B \ + \ H-H \ \longrightarrow \ HA \ + \ BH \qquad (2)$$

$$A,B \ = \ C,N,O,S,Hal.$$

It may be noted that reactions of this type are symmetry-forbidden in the ground state as far as suprafacial reaction pathways are concerned[1]. The majority of the transition elements possesses at least some catalytic activity with regard to these reactions, but we will be concerned mainly with the metals of the platinum group[2]:

	Fe	Co	Ni	(Cu)
	Ru	Rh	Pd	
(Re)	Os	Ir	Pt	

Broadly speaking, catalysts come in two physical forms: (micro)crystalline and atomically dispersed. This latter group presents the simplest mechanistic picture, since each single metal atom serves as a focus of catalytic activity, and participates in the reactions without interaction with the other metal atoms. Crystalline catalysts are somewhat more complicated[3]. The interaction of the catalyst surface with the reactants is conveniently described in terms of *active sites*, each comprising a number of metal atoms. The dilution of catalytically active metal atoms by alloying constitutes a recent development[4], which may become of great importance for both mechanistic and synthetic purposes.
The reaction pattern for the majority of the metal catalyzed hydrogenations

3

can be summarized as follows: the π-bond and hydrogen form coordinative bonds with the active site, followed by subsequent transfer of the hydrogen atoms. The reaction scheme is depicted below for (Z)-2-butene. The transfer of

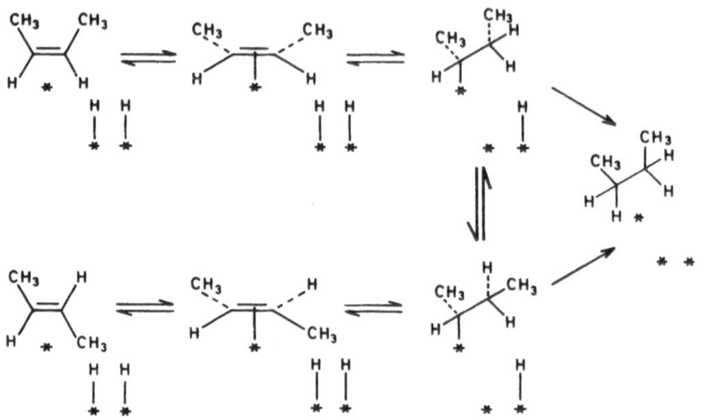

the first hydrogen atom is reversible: if another hydrogen atom than the one which had originally been transferred is abstracted, isomerization of the double bond results.

A reaction scheme for hydrogenolysis can be depicted as follows:

A temporary bond between reactant and catalyst may also be assumed in this case, an unsaturated function at the α- or β-position may serve as a 'handle'.

References

1. R.G. Pearson, *Chem. Eng. News*, **1970** (Sept. 28), 66.
2. Further notable omissions are: catalysis by alkali metal naphthalides and related systems, *cf.* K. Tamaru, *Adv. Catal.*, **20**, 327 (1969); ionic hydrogenation, D.N. Kursanov, Z.W. Parnes, and N.M. Loim, *Synthesis*, **1974**, 633; bacterially catalyzed hydrogenation, H. Simon, B. Rambeck, H. Hashimoto, H. Günter, G. Nohynek, and H. Neumann, *Angew. Chem.*, **86**, 675 (1974); B. Rambeck and H. Simon, *Ibid.*, **86**, 675 (1974).
3. V.A. Dzis'ko, *Russ. Chem. Rev.* (Engl. Transl.), **43**, 435 (1974).
4. J.K.A. Clarke, *Chem. Rev.*, **75**, 291 (1975); V. Ponec, *Catal. Rev.*, **11**, 41 (1975).

2. The active site

Catalytic hydrogenation and hydrogenolysis involve covalent bonding between the reactants and the catalyst (chemisorption). The number of metal atoms available for bonding constitutes a fundamental difference between heterogeneous and homogeneous catalysts[1]. In the latter case, each active site consists of one metal atom which should be able to accomodate the reactants in its coordination sphere. For that reason, coordinative unsaturation and the presence of at least one labile ligand are characteristic traits of homogeneous catalysts.

With heterogeneous catalysts more metal atoms may be involved in the reaction[2]. A low coordination number remains a necessity, however, and accounts for the observation that only a small fraction of the metal surface contributes to the overall activity[3]. According to the present theory, active sites for structure-sensitive ('demanding') reactions[4] are situated at corners, edges, and crystal defects[5], where the coordination number of the surface atom is lower than at regular surface planes[6]. Hydrogenolysis and isomerization seem to be structure-sensitive reactions; whether this is also true for hydrogenation is not yet clear.

The nature of the chemisorptive bond[7] has been and still is a subject of investigation. Homogeneous catalysis can obviously be described in terms of coordinative bonds according to the Chatt-Dewar-Duncanson model[8],

Orbitals used in the combination of ethylene with platinum.

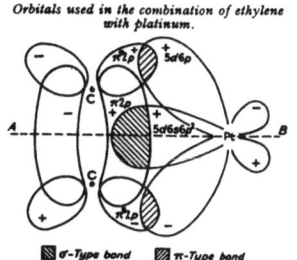

◼ σ-Type bond   ▨ π-Type bond

Spatial arrangement of atoms in $[C_2H_4PtCl_3]^-$.

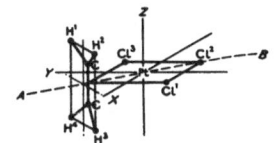

(The plane of the hydrogen atoms is parallel to the plane Cl^1ZCl^2 but probably displaced slightly from co-planarity with the carbon atoms by the repulsion of the hydrogen atoms by the platinum atom.)

(Figure reprinted from: J. Chatt and L.A. Duncanson, *J. Chem. Soc.*, 1953, 2939 by permission of the copyright holder)

which also serves as a model for chemisorption.[9] It has been stated[10] that even larger molecules like benzene are chemisorbed on a single metal atom, which would make the analogy with coordination chemistry complete.

As regards heterogeneous hydrogenation, it has already been pointed out that the reactants are to share an active site; only those locations at the catalyst surface which can accomodate hydrogen as well as organic reactant exhibit catalytic activity. The rate of reaction is proportional to the number of sites (and therefore to the amount of catalyst). For the heterogeneously catalyzed liquid-phase reaction the reaction rate of a compound A (r_A) may be expressed, using Langmuir-Hinshelwood kinetics[11], by

$$r_A = -\frac{d[A]}{dt} = k_A \, \Theta_A \, w \, (p\text{-}p_s)^n = \frac{k_A b_A [A] w (p\text{-}p_s)^n}{1 + b_A [A] + \Sigma bc}$$

in which b_A is the adsorption constant, k_A is the reaction rate constant, Θ_A is the fraction of the active catalyst surface covered by A, w is the amount of catalyst, p is the pressure, p_s is the vapour pressure of the solvent and Σbc is the sum of the contributions of the solvent, the hydrogenated product, and the hydrogen to the denominator of the Langmuir expression (due to partial coverage of the catalyst by these compounds)[12]. Depending on the reaction conditions and catalyst, the order in hydrogen (n) may vary from 0 to 1. At a constant hydrogen pressure, we may write

$$r'_A = k'_A \Theta_A$$

in which r'_A and k'_A are the reaction rate and pseudo reaction rate constant per weight amount of catalyst. In most hydrogenations the reactant covers the catalyst surface almost completely ($\Theta_A \approx 1$)[12,13], whilst the chemisorbed hydrogen usually does not affect the chemisorption of the reactant. On the other hand, hydrogenolysis reactions are seriously retarded by the product formed[14]. This is due to comparable strengths of adsorption of reactant and product, so that Θ_A decreases as the reaction proceeds[15].

When two compounds A and B are hydrogenated in competition, a chemisorption equilibrium

$$A + \overset{B}{\underset{*}{|}} \rightleftharpoons \overset{A}{\underset{*}{|}} + B$$

is involved which can be described by[12]

$$K_{A,B} = \frac{b_A}{b_B} = \frac{\Theta_A [B]}{[A] \Theta_B}$$

6

and

$$r'_A/r'_B = \frac{d[A]}{d[B]} = \frac{k'_A \Theta_A}{k'_B \Theta_B} = \frac{k'_A}{k'_B} K_{A,B} \frac{[A]}{[B]}$$

Consequently, the selectivity is dependent on both the rate constants k' and the position of the chemisorption equilibrium[12,15] (relative strength of adsorption).

For consecutive reactions of the type

$$A \longrightarrow B \longrightarrow C$$

it is often desired to obtain B in high yield. Here, the overall rate of formation of B is given by

$$r'_B = \frac{d[B]}{dt} = k'_A \Theta_A - k'_B \Theta_B$$

The highest concentration of B is reached when $d[B]/dt = 0$, *i.e.* $k'_A \Theta_A = k'_B \Theta_B$.

Therefore, a favourable ratio $[B]/[A] = (k'_A/k'_B)K_{AB}$ will be obtained if A reacts more rapidly and/or adsorbs more strongly than B[17].

As an illustration the selectivity of such a reaction towards B has been depicted below with $b_A = 1$, $k'_A = 2$, $k'_B = 1$, and $b_B = 10$, 1, and 0.1, respectively[17].

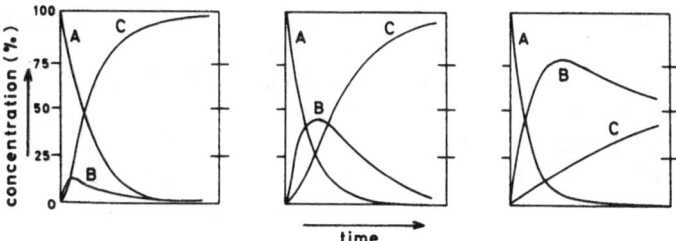

Finally, it should be noted that in the foregoing consideration it has been assumed that neither adsorption nor desorption of the reactants and products is rate-determining, *i.e.* the surface transformation of A and B is the slow step of the reaction process. In most instances, this assumption has been found to be correct.

These considerations are crucial when selective hydrogenation is the object, *e.g.* partial hydrogenation of acetylenes (p 62) or conjugated or skipped polyenes (p 45). In such cases, the rate constants are of similar magnitude and the selectivity depends solely on the position of the chemisorption equilibrium.

7

As regards homogeneous hydrogenation, the position of the coordination equilibria and the rate of the rearrangement steps decide the kinetics and selectivity of the reaction. Because of the widely diverging reaction schemes[18] it is impossible to give a unified treatment.

References

1. This consideration has important consequences for the selectivity of the reactions, *vide infra.*
2. *Cf.* P. Ratnasamy, *J. Catal.*, **31**, 466 (1973).
3. I. Langmuir, *Trans. Faraday Soc.*, **17**, 607 (1922).
4. M. Boudart, A. Aldag, J.E. Benson, N.A. Dougharty, and C.G. Harkins, *J. Catal.*, **6**, 92 (1966).
5. G.C. Bond, in *Mechanism of Hydrocarbon Reactions,* F. Marta and D. Kalls, Eds., Elsevier, Amsterdam, 1975, p 49.
6. For a face centered cubic metal, the coordination number of a surface atom is 8 for an atom in a (100) plane, 7 for a (110) plane and 9 for a (111) plane, whereas for a bulk atom it is 12.
7. See A. Clark, *The chemisorptive bond,* Academic Press, New York, 1974.
8. M.J.S. Dewar, *Bull. Soc. Chim. Fr.*, **18**, C79 (1951); J. Chatt and L.A. Duncanson, *J. Chem. Soc.*, 2939 (1953); see also: F.R. Hartley, *Chem. Rev.*, **69**, 799 (1969).
9. G.C. Bond, *Discuss. Faraday Soc.*, **41**, 200 (1966).
10. J.M. Basset, G. Dalmai-Imelik, M. Primet, and R. Mutin, *J. Catal.*, **37**, 22 (1975), and pertinent references.
11. I. Langmuir, *J. Am. Chem. Soc.*, **38**, 2221 (1916); C.N. Hinshelwood, *Annu. Rep. Chem. Soc., London,* **27**, 11 (1930).
12. *Cf.* A.P.G. Kieboom and H. van Bekkum, *J. Catal.*, **25**, 342 (1972).
13. *Cf.* H. van Bekkum, A.P.G. Kieboom, and K.J.G. van de Putte, *Rec. Trav. Chim. Pays-Bas,* **88**, 52 (1969).
14. R.W. Meschke and W.H. Hartung, *J. Org. Chem.*, **25**, 137 (1960).
15. A.P.G. Kieboom, J.F. de Kreuk, and H. van Bekkum, *J. Catal.*, **20**, 58 (1971).
16. A.S. Hussey, R.H. Baker, and G.W. Keulks, *Ibid.,* **10**, 258 (1968); Y. Moro-Oka, T. Kitamura, and A. Ozaki, *Ibid.,* **13**, 53 (1969).
17. L. Beránek, *Adv. Catal.*, **24**, 1 (1975), and pertinent references.
18. B.R. James, *Homogeneous Hydrogenation,* Wiley-Interscience, London, 1973.

3. Mechanistic investigations

Extensive mechanistic research has been carried out on both heterogeneously and homogeneously catalyzed hydrogenation and hydrogenolysis in order to explain the numerous phenomena connected with these reactions. One of the major purposes of these mechanistic studies is a better understanding of the reaction pattern, from which it would be possible to predict to some extent the course of related conversions. The more relevant results of these investigations will therefore be dealt with alongside the respective hydrogenation and hydrogenolysis reactions, together with their practical synthetic implications. In addition, some general aspects have been summarized below.

Homogeneously catalyzed hydrogenation is usually discussed in terms of successive transfer of two hydrogen atoms to the coordinated reactant. *Cis*-ligand insertion (or more appropriately, *cis*-ligand migration), which is a very common reaction in coordination chemistry[1], serves as a model for the transfer of the first hydrogen atom, with the formation of a σ-bonded complex. Transfer of the second hydrogen atom then gives the hydrogenated product. The second hydrogen atom is usually transferred from the catalyst *via* a three-centre transition state, but solvent participation has also been demonstrated[2]. As regards the steric course of the reaction, the *cis*-migration is often viewed as a suprafacial process, although antarafacial reactions have been reported[3,4]. The transfer of the second hydrogen atom proceeds suprafacially if the metal atom is involved, antarafacially in the case of solvolysis of the σ-bond[2].
Deuterium labelling as well as the study of steric pathways has been important in the elucidation of reaction mechanisms. Moreover, the key steps have been studied separately with suitable coordination compounds.

In the field of *heterogeneous* catalysis the active site of the catalyst cannot be defined precisely. Mechanistic studies have been carried out in two different ways:
i. testing a series of metals by means of a standard reaction on just one reactant[5],
ii. studying the reaction of a series of compounds on a single metal[6].

Both methods may give a picture of electronic as well as steric interactions between the reactant and the metal surface, together with the possible effects thereon of the solvent. Numerous studies concerning stereochemistry (in particular C=C hydrogenation[7] and C-C, C-O, and C-N hydrogenolysis[8]), the influence of steric and electronic effects on both the rate of reaction and the strength of adsorption[9] (e.g. by means of linear free energy relationships[6]), and the influence of the solvent[10] have led to the following conclusions:

Hydrogen is dissociatively chemisorbed on the catalyst and the hydrogenation or hydrogenolysis reaction has to be considered as a stepwise addition of two hydrogen atoms from the catalyst metal to the reactant (Horiuti and Polanyi mechanism[11]). In some cases one hydrogen may be delivered by the solvent in the form of a proton, while simultaneously a neighbouring hydrogen atom adsorbed on the catalyst goes into solution as a proton[12]. There are some indications that the electronic nature of the hydrogen species which attacks the reactant is dependent on the reactant molecule involved. While several hydrogenolyses may be best described in terms of a hydride attack (S_N1, S_N2 and S_Ni mechanism), most hydrogenations point rather to atomic hydrogen as the reducing agent.

References

1. J.P. Collman, *Acc. Chem. Res.*, **1**, 136 (1968); see also P. Cossee, *Rec. Trav. Chim. Pays-Bas*, **85**, 1151 (1966).
2. H. van Bekkum, F. van Rantwijk, G. van Minnen-Pathuis, J.D. Remijnse, and A. van Veen, *Rec. Trav. Chim. Pays-Bas*, **88**, 911 (1969).
3. J. Ashley-Smith, M. Green, and P.C. Wood, *J. Chem. Soc. A*, **1970**, 1847; D. Hudson, D.E. Webster, and P.B. Wells, *J. Chem. Soc., Dalton Trans.*, **1972**, 1204.
4. F. van Rantwijk and H. van Bekkum, *J. Mol. Catal.*, **1**, 383 (1975/76).
5. M. Boudart, *Chem. Eng. Prog.*, **57** (8), 33 (1961).
6. M. Kraus, *Adv. Catal.*, **17**, 75 (1967); I. Mochida and Y. Yoneda, *J. Catal.*, **11**, 183 (1968); A.P.G. Kieboom, *Substituent effects in the hydrogenation on palladium*, Thesis, Delft, 1971; A.P.G. Kieboom, *Bull. Chem. Soc. Jpn.*, **49**, 331 (1976).
7. S. Siegel, *Adv. Catal.*, **16**, 123 (1966).
8. A.P.G. Kieboom, A.J. Breijer, and H. van Bekkum, *Rec. Trav. Chim. Pays-Bas*, **93**, 186 (1974); S. Mitsui, S. Imaizumi, and Y. Esashi, *Bull. Chem. Soc. Jpn.*, **43**, 2143 (1970); Y. Sugi and S. Mitsui, *Ibid.*, **43**, 1489 (1970), and references cited in these papers.
9. I. Jardine and F.J. McQuillin, *Tetrahedron Lett.*, **1968**, 5189; A.S. Hussey and G.P. Nowack, *J. Org. Chem.*, **34**, 439 (1969); C.P. Rader and H.A. Smith, *J. Am. Chem. Soc.*, **84**, 1443 (1962); J. Völter, M. Hermann, and K. Heise, *J. Catal.*, **12**, 307 (1968), and references cited in these papers.

10. L. Cerveny, A. Prochazka, and V. Ruzicka, *Coll. Czech. Chem. Commun.*, **39**, 2463 (1974), and references.
11. I. Horiuti and M. Polanyi, *Trans. Faraday Soc.*, **30**, 1164 (1934).
12. F. van Rantwijk, A. van Vliet, and H. van Bekkum, *J. Chem. Soc., Chem. Commun.*, **1973**, 234; A.P.G. Kieboom, J.F. de Kreuk, and H. van Bekkum, *J. Catal.*, **20**, 58 (1971); A.P.G. Kieboom, A.J. Breijer, and H. van Bekkum, *Rec. Trav. Chim. Pays-Bas*, **93**, 186 (1974); F. van Rantwijk, A.P.G. Kieboom, and H. van Bekkum, *J. Mol. Catal.*, **1**, 27 (1975/76); A.P.G. Kieboom, H.J. van Benschop, and H. van Bekkum, *Rec. Trav. Chim. Pays-Bas*, **95**, 231 (1976).

4. The catalyst

With the exception of the three lightest ones, the group VIII metals are rather expensive. In order to obtain the maximum number of active sites per unit weight of metal, the metal should be highly disperse (dispersion = surface atoms per total metal atoms). The dispersion of the metal can be stabilized against aggregation in a number of ways.

The oldest method is colloidal suspension but this has now been superseded. Application of the metal as a highly porous solid (metal black) is still used, especially with the less expensive metals. Dispersion of microcrystalline metal particles on a solid support has many advantages and is widely applied.

In order to avoid the problems connected with a reaction at a solid/liquid inter-face, homogeneous catalysts have been developed in recent years; these have the concomitant advantage that each metal atom is available for reaction. In these catalysts, the metal is present as a low-valent ion, stabilized by suitable ligands against reduction to the metal. The constitution of homogeneous catalysts has to strike a fine balance between stability against oxidation and reduction on one hand and a low coordination number − in order to accomo-date the reactants − on the other.

With homogeneous catalysts, the separation of the catalyst from the reaction products, preferably combined with re-use of the catalyst, presents serious technological problems. Heterogenization of homogeneous catalysts, by attach-ment to an insoluble support, is a promising development, which may combine the favourable characteristics of supported heterogeneous and homogeneous catalysts.

HETEROGENEOUS CATALYSTS

The older catalysts consisted of the pure metal (known as metal black), either in the form of a colloid or in the form of a suspension. Among these, platinum and, to a somewhat lesser extent, palladium have been widely used as suspen-sion catalysts. They are prepared *in situ* by reduction of the corresponding (specially prepared) metal oxides with hydrogen. At present, platinum is still often used in this way (Adams catalyst)[1]. Nickel is most frequently applied as a suspension catalyst (Raney nickel),[2] since it is easily prepared by treatment of the inexpensive nickel-aluminium alloy with aqueous alkali. In this way a highly active nickel sponge is obtained (surface area about 80 m^2/g, mean pore diameter about 60 Å). In addition, nickel (and also cobalt, copper and iron) suspensions may be obtained by treating aqueous solutions of their salts with zinc or aluminium, followed by digestion of the precipitated metal with alkali or acid (Urushibara catalysts)[3]. More recently, transition metal borides have found application (both as suspension and as supported metal catalysts)[4]. These useful catalysts are easily obtained by sodium borohydride reduction of transition metal salts. The exact chemical nature of the metal borides thus obtained is, however, rather complex and is not completely established[5].

Supported metal catalysts are most frequently used since then the metal is present in a more dispersed form compared with suspension catalysts. The application of a support with a high surface area allows the occurrence of very small stable metal crystallites (bound to the support), resulting in a much higher specific activity of the metal (i.e. number of active sites per unit weight of metal). Furthermore, the solid support improves the stability of the catalyst. Carbon (surface area 500–1000 m^2/g), silica (100–300 m^2/g), and alumina (75-350 m^2/g) are the supports commonly used. In addition, inorganic metal salts (e.g. calcium carbonate, barium sulfate), molecular sieves, carbon molecular sieves, and organic polymeric materials have been applied in some cases[6]. The advantage of a solid support may be illustrated by the fact that palladium black has a surface area of 5–10 m^2/g Pd whereas for 10% palladium on carbon this value is 100–200 m^2/g Pd. An increase of the metal loading of the supported catalyst leads to the formation of larger crystallites. The active surface area (number of active sites) per unit weight of metal diminishes for higher metal loadings (see Table I).

Table I: The active surface area per unit weight of metal

%Pd (w/w on alumina)	Metal surface area[a] (m^2/g Pd)
0.32	90
0.86	70
1.00	62
1.37	57
1.82	51
4.21	36

a: Average values from ref [7].

As a consequence, the specific catalytic activity (defined as the reaction rate

13

per unit weight of metal) decreases as demonstrated in Table II for the hydrogenation of cyclohexene over various alumina-supported platinum catalysts[8].

Table II: The hydrogenation of cyclohexene over Pt/Al$_2$O$_3$

% Pt (w/w on alumina[b])	Specific Rate[a] (mole min^{-1} gat^{-1} atm^{-1})
0.39	121
0.49	117
0.52	113
0.64	104
0.72	96
0.97	81
1.11	76

a: Hydrogenation of cyclohexene in cyclohexane at 25° and 1 atm.
b: Surface area 190 m^2/g.

It may be noted that this effect in this loading region will be smaller if the surface area of the support is higher (e.g. carbon with a surface area > 500 m^2/g).

The use of heterogeneous catalysts introduces very easily diffusion limitation in the transport of reactants and products between the solution and the metal[9]. In particular, the smaller pores may cause serious transport limitations. Consequently, apart from other effects, the use of different catalyst supports might result in some change in product distribution (selectivity). For example, the selectivity of the hydrogenation of polyunsaturated triglycerides was found to be strongly dependent on the pore size distribution of the silica used as the catalyst support[10].

The supported transition metal catalysts may be prepared very easily. After impregnation or adsorption of the metal salt (in aqueous solution) in or on the solid support, the metal is reduced by hydrogen, formaldehyde or sodium borohydride. Subsequently, the catalyst is filtered, thoroughly washed and dried. The various procedures have been described in detail in the literature[11,12]. In addition, most of the catalysts are commercially available and are ready for use without any pre-activation. A comprehensive survey of the various techniques which are available for catalyst characterization is given below. A detailed review of these measurement techniques was given recently by Anderson[13].

Information on the *total surface area* of the catalyst (metal plus support) can be obtained by the classical BET method of low temperature physical adsorption of an inert gas (nitrogen or one of the rare gases)[14,15]. From the physical adsorption isotherm (often measured at the temperature of boiling nitrogen, 77 K) the amount of gas adsorbed in a monolayer can be determined. This, together with the surface area occupied by one gas molecule, gives directly the total surface area of the catalyst. Surface areas between a few and 1500 m^{2}g^{-1} can be determined.

14

Insight into the *pore size distribution* of the catalyst can be deduced from both mercury penetration measurements[16] and/or determination of the complete gas adsorption-desorption isotherm[15]. The latter method consists merely of more sophisticated BET measurements, which include the determination of the adsorption-isotherm up to the saturation vapour pressure and the reverse process (desorption-isotherm). The mercury penetration is based on the capillary depression of mercury (due to its high cohesion with respect to its adhesion to most other materials). The smaller the pore diameter, the higher becomes the pressure needed to fill the pores with mercury (100 A needs 750 atm, 25 A needs 3000 atm pressure). Volume-pressure relations for both methods enable the determination of the pore size distribution.

The *metal dispersion* (surface area and/or particle size) can be determined by different methods:[14,17]

, i. Chemisorption methods using hydrogen or carbon monoxide as the chemisorbate[14,18]. The selective chemisorption of the gas on the metal gives directly the number of surface atoms of the metal from the amount of gas chemisorbed. Of course, one has to know the stoichiometry of the chemisorption process (*e.g.* for Pt-CO this is 1:1).

ii. X-ray diffraction measurements give, from the broadening of the diffraction lines, an estimate of the average crystallite size for crystallites of $30-1000$ Å[19]. In addition, the difference in intensities of the diffraction lines before and after sintering of the catalyst allows an estimation of the percentage of crystallites <50 Å

iii. Electron microscopic determination of the metal crystallite size (> 10 A) requires a number of measurements in order to get a reliable picture[20] due to the very small probes applied.

Temperature programmed desorption (TPD) of chemisorbed hydrogen gives a distribution of the various active sites (and their surface concentration) of the catalyst metal[21].

Finally, a more detailed description of the *surface structure* of metallic catalysts may be obtained by physical analysis techniques[13,22] such as low-energy electron diffraction (LEED), ion scattering spectroscopy (ISS), Auger electron spectroscopy (AES), secondary ion mass spectrometry (SIMS), bombardment light emission (BLE), and ESCA techniques like X-ray photoelectron spectroscopy (XPS) and ultraviolet photoelectron spectroscopy (UPS).

HOMOGENEOUS CATALYSTS

Homogeneous catalysts avoid the transport problems of supported heterogeneous catalysts. Moreover, each metal atom is available for reaction. An active homogeneous catalyst represents a subtle balance of stabilization — prior to use as well as in the course of the reaction — by suitable ligands on one hand and preservation of at least one vacant coordination site on the other. A number of ready-for-use catalysts — mostly low-valent ruthenium and rhodium phosphine complexes — have been developed in recent years and are commercially available. The complications resulting from the inherent instability of catalytically active species, especially towards oxidation, can be circumvented by preparation of the catalyst *in situ* from a stable precursor. It should be noted that the number of synthetically useful homogeneous catalysts is rather small. A somewhat larger number have been used in mechanistic studies whilst many metal complexes exhibit reaction patterns related to homogeneous catalysis.

With homogeneous catalysts an active site is limited to a single metal atom.

The number of available positions for coordination are therefore limited when compared with heterogeneous catalysts. For that reason, homogeneously catalyzed arene hydrogenation has been notably difficult to achieve[23], whilst the ability of homogeneous catalysts to effect hydrogenolysis is negligible and is limited to vinyl and allyl compounds[24].

Up to the present, the application of homogeneous catalysts to organic synthesis has been limited[25]. The rather low activity of homogeneous catalysts may play a role in this. On the other hand chiral homogeneous catalysts have proved very useful for enantioselective hydrogenation of prochiral alkenes, especially amino acid precursors. Indeed, homogeneous catalysts represent a breakthrough in the field of enantioselective hydrogenation, a breakthrough which has not been forthcoming with heterogeneous catalysts[26].

A disadvantage of soluble catalysts is the difficulty of separating the catalyst and the reaction products. In addition to the use of supported metal complexes[27] (see below) the use of soluble macromolecular metal complexes has been proposed[28]; these may be easily separated from the reaction mixture by gel or membrane filtration. An example is a soluble rhodium triphenylphosphine complex attached to linear polystyrene[28].

IMMOBILIZED HOMOGENEOUS CATALYSTS

Immobilized homogeneous catalysts have been developed recently in order to circumvent the inherent practical complications of separating soluble catalysts. These 'heterogenized' homogeneous catalysts should in principle combine the favourable characteristics of homogeneous and heterogeneous catalysts[27].

Homogeneous catalysts have been immobilized by attachment to a polymeric ligand (mostly modified polystyrene-divinylbenzene)[29] and also by covalent bonding to silica. In this latter field employment of a swelling layer lattice silicate is a recent development[30].

With a number of model systems, using the polystyrene-divinylbenzene support, the stability as well as the easy separation and re-use of immobilized homogeneous catalysts has been demonstrated[31]. In some cases, a catalyst is obtained whose hydrogenation efficiency is greater than the corresponding nonattached metal complex, because of prevention of dimerization of the metal complex[32]. In particular, immobilization of the very expensive chiral homogeneous catalysts is of great value.[33] However, the diffusion problems of all supported catalysts manifest themselves. The choice of solvent is rather limited, since swelling of the support is a necessary requirement. Depending on the properties of the support, immobilized catalysts sometimes exhibit low reactivity towards bulky substrates[34].

The technique of catalyst immobilization evidently needs ideally a support which is able to swell in a wider variety of solvents and which is rather loosely cross-linked in order to present wide mazes.

References

1. O. Loew, *Ber. Dtsch. Chem. Ges.* **23**, 289 (1890); V. Voorhees and R. Adams, *J. Am. Chem. Soc.*, **44**, 1397 (1922); R. Adams and R.L. Shriner, *Ibid.*, **45**, 2171 (1923); R. Adams, V. Voorhees, and R.L. Shriner, *Org. Synth.*, Coll. Vol. 1, 463 (1932).

2. M. Raney, U.S. Patents 1,563,687 (1927), 1,628,190 (1927), and 1,915,473 (1933); R. Mozingo, *Org. Synth.*, Coll. Vol. 3, 181 (1955); A.A. Pavlic and H. Adkins, *J. Am. Chem. Soc.*, **68**, 1471 (1946); H. Adkins and A.A. Pavlic, *Ibid.*, **69**, 3039 (1947); H. Adkins and H.R. Billica, *Ibid.*, **70**, 695 (1948); X.A. Dominguez, I.C. Lopez, and R. Franco, *J. Org. Chem.*, **26**, 1625 (1961).

3. Y. Urushibara, *Ann. N.Y. Acad. Sci.*, **145**, 52 (1967).

4. H.I. Schlesinger, H.C. Brown, A.E. Finholt, J.R. Gilbreath, H.R. Hoekstra, and E.K. Hyde, *J. Am. Chem. Soc.*, **75**, 215 (1953); H.C. Brown and C.A. Brown, *Ibid.*, **84**, 1495 (1962); C.A. Brown and H.C. Brown, *J. Org. Chem.*, **31**, 3989 (1966); C.A. Brown, *J. Org. Chem.*, **35**, 1900 (1970); C.A. Brown and V.K. Ahuja, *Ibid.*, **38**, 2226 (1973).

5. P.C. Maybury, R.W. Mitchell, and M.F. Hawthorne, *J. Chem. Soc., Chem. Commun.*, **1974**, 534, and references cited therein.

6. *Cf.* J. Sabadie and J.E. Germain, *Bull. Soc. Chim. Fr.*, **1974**, 1133; R.L. Lazcano, M.P. Pedrosa, J. Sabadie, and J.E. Germain, *Ibid.*, **1974**, 1129; O.A. Tyurenkova, *Russ. J. Phys. Chem.* (Engl. Transl.), **43**, 1167 (1969); D.L. Trimm and B.J. Cooper, *J. Catal.*, **31**, 287 (1973); H. van Bekkum and D.P. Roelofsen, *Chem. Tech. (Amsterdam)*, 28, 249 (1973).

7. T. Paryjczak and K. Jozwiak, *J. Chromatogr.*, **111**, 443 (1975).

8. A.S. Hussey, G.W. Keulks, G.P. Nowack, and R.H. Baker, *J. Org. Chem.*, **33**, 610 (1968).

9. G.J.K. Acres and B.J. Cooper, *J. Appl. Chem. Biotechnol.*, **22**, 769 (1972); F. Nagy, A. Petho, and D. Moger, *J. Catal.*, **5**, 348 (1966).

10. C. Okkerse, A. de Jonge, J.W.E. Coenen, and A. Rozendaal, *J. Am. Oil Chem. Soc.*, **44**, 152 (1967); C. Okkerse, *Chem. Weekbl.*, **63**, 237 (1967).

11. R.L. Augustine, *Catalytic Hydrogenation*, Marcel Dekker, New York, 1965, pp 147-153; R. Mozingo, *Org. Synth.*, Coll. Vol. 3, 685 (1955); A.I. Vogel, *Practical Organic Chemistry*, 3rd ed., Longmans, London, 1972, pp 699, 948-951; F. Zymalkowski, *Katalytische Hydrierungen*, Enke Verlag, Stuttgart, 1965, pp 23-34.

12. J.R. Anderson, *Structure of Metallic Catalysts*, Academic Press, London, 1975, pp 451-459.

13. Reference 12, pp 289-444.

14. T.E. Whyte, Jr., *Catal. Rev.*, **8**, 117 (1973).

15. J.C.P. Broekhoff and B.G. Linsen, in *Physical and Chemical Aspects of Adsorbents and Catalysts*, B.G. Linsen, Ed., Academic Press, London, 1970, pp 1-62.

16. H.L. Ritter and L.C. Drake, *Ind. Eng. Chem. Anal. Ed.*, **17**, 782 (1945); L.C. Drake and H.L. Ritter, *Ibid.*, **17**, 787 (1945).

17. *Cf.* P.A. Sermon, *J. Catal.*, **24**, 467 (1972), and references cited herein.

18. R.J. Farranto, *Chem. Eng. Prog.*, **71**, 37 (1975); J.J.F. Scholten and A. van Montfoort, *J. Catal.*, **1**, 85 (1962); C.E. Hunt, *Ibid.*, **23**, 93 (1971); P.A. Sermon, *Ibid.*, **24**, 460 (1972).

19. *Cf.* D. Pope, W.L. Smith, M.J. Eastlake, and R.L. Moss, *J. Catal.*, **22**, 72 (1971), and references.

20. P.C. Flynn, S.E. Wanke, and P.S. Turner, *J. Catal.*, **33**, 233 (1974), and references cited herein.

21. P.C. Aben, H. van der Eijk, and J.M. Oelderik, *Proc. 5th Int. Congr. Catal., 1972*, **1**, 717 (1973).

22. *Chem. Weekbl.*, **71** (12), 13-28 (1975).

23. Recently, E.L. Muetterties and F.J. Hirsekorn, *J. Am. Chem. Soc.*, **96**, 4063 (1974), reported the first unequivocal demonstration of homogeneous benzene hydrogenation with a metal complex.

24. Pp 92, 125.

25. P.N. Rylander and L. Hasbrouck, *Engelhard Ind. Tech. Bull.*, **10**, 85 (1969); F.J. McQuillin, *Prog. Org. Chem.*, **8**, 314 (1973); B.R. James, *Homogeneous Hydrogenation*, John Wiley, New York, 1973.

26. Pp 80–81.

27. J. Manassen, *Platinum Met. Rev.*, **15**, 142 (1971); Z.M. Michalska and D.E. Webster, *Platinum Met. Rev.*, **18**, 65 (1974); J.C. Bailar, Jr., *Catal. Rev.*, **10**, 17 (1974); see also *Solid-Phase Synthesis*, E.C. Blossey and D.C. Neckers, Eds., Halsted Press, Stroudsburg, 1975, pp 284-292, and J.I. Crowley and H. Rapoport, *Acc. Chem. Res.* **9**, 135 (1976).

28. E. Bayer and V. Schurig, *Angew. Chem.*, **87**, 484 (1975).

29. *Cf.* J.P. Collman, L.S. Hegedus, M.P. Cooke, J.R. Norton, G. Dolcetti, and D.N. Marquardt, *J. Am. Chem. Soc.*, **94**, 1789 (1972); K.G. Allum, R.D. Hancock, I.V. Howell, R.C. Pitkethly, and P.J. Robinson, *J. Organomet. Chem.*, **87**, 189 (1975); C.U. Pittman, Jr., B.T. Kim, and W.M. Douglas, *J. Org. Chem.*, **40**, 590 (1975).

30. T.J. Pinnavaia and P.K. Welty, *J. Am. Chem. Soc.*, **97**, 3819 (1975).

31. C.U. Pittman, Jr., L.R. Smith, and R.M. Hanes, *Ibid.*, **97**, 1742 (1975).

32. W.D. Bonds, Jr., C.H. Brubaker, Jr., E.S. Chandrasekaran, C. Gibbons, R.H. Grubbs, and L.C. Kroll, *J. Am. Chem. Soc.*, **97**, 2128 (1975); C.U. Pittman, Jr., S.E. Jacobson, and H. Hiramoto, *Ibid.*, **97**, 4774 (1975).

33. W. Dumont, J.C. Poulin, T.P. Dang, and H.B. Kagan, *J. Am. Chem. Soc.*, **95**, 8295 (1973).

34. R.H. Grubbs, L.C. Kroll, and E.M. Sweet, *J. Macromol. Sci., Chem. A*, 7, 1047 (1973).

5. The Metals

Nickel has been applied in many cases for both hydrogenation and hydrogenolysis reactions since Sabatier's original experiments[1]. For laboratory use, the metal is used in the form of Raney nickel[2], which is of particular importance for carbon-sulphur hydrogenolysis (desulphurization). Supported nickel catalysts are used in industry. More recent developments are the so-called P-1 and P-2 nickel[3], prepared by sodium borohydride reduction of nickel salts.

Cobalt has been used as Raney cobalt for the hydrogenation of nitriles[4]. Potassium pentacyanocobaltate[5] is the oldest homogeneous hydrogenation catalyst. Recently, an active cobalt complex $(Co(\eta^3-C_3H_5)[P(OCH_3)_3]_3)$ has been developed for benzene hydrogenation[6].

Copper has been frequently applied as copper chromite (Adkins catalyst)[7] to the selective hydrogenation of carbonyl functions. The rather drastic reaction conditions limit its use in the laboratory, especially in view of the numerous mild metal hydride reduction procedures available[8]. A recent industrial development is a chromite catalyst for the selective hydrogenolysis of glycerol trilinoleate[9].

Platinum is applied as a dispersion on carbon or alumina or as a suspension (Adams catalyst)[10]. Platinum is a highly active catalyst for the hydrogenation of alkenes, but is not very selective. Double bond isomerization and hydrogenolysis do not occur readily. Furthermore, it promotes the hydrogenation of most other functional groups under fairly mild conditions. The platinum-tin chloride complex is a useful homogeneous catalyst[11].

Palladium is used in a similar way to platinum. It has a very high hydrogenolytic activity. Its activity for alkene hydrogenation is somewhat lower than that of platinum and considerable double bond migration occurs. Almost no conversion of aliphatic ketones and aromatic compounds takes place at ambient conditions. Poisoned palladium catalysts find extensive application for the selective semi-hydrogenation of alkynes[12], *e.g.* the Lindlar catalyst.

Rhodium is used as a dispersion on carbon or alumina and has a high activity for the hydrogenation of aromatic compounds. Its activity for alkene hydrogenation is rather low compared with platinum or palladium. In some cases the selectivity is much better. Chlorotris(triphenylphosphine)rhodium(I) (Wilkinson catalyst) is the most important homogeneous catalyst[13].

Ruthenium on carbon is applied to the hydrogenation of aromatic rings and carbonyl functions. It has the lowest hydrogenolytic activity of the transition metals, and is applicable when high selectivity is required. Chlorohydridotris-(triphenylphosphine)ruthenium(II) is a homogeneous catalyst which is suitable for the hydrogenation of terminal alkenes.

Iridium, osmium and *iron* are of minor importance as hydrogenation catalysts.

Rhenium has been used for the selective hydrogenolysis of carboxylic acids to alcohols[14].

Gold has very recently been found to catalyze the hydrogenation of some simple alkenes[15].

Transition metal sulphides[16] require rather drastic reaction conditions and are of technical importance for the hydrogenation of sulphur-containing compounds.

Recently various *alloys* have been investigated (*e.g.* Pt-Rh, Pt-Pd, Pd-Ag, Pd-Au Ni-Cu) as catalyst systems[17]. Synthetic applications are promising for conversions of compounds with several functional groups in which both hydrogenation and hydrogenolysis can occur at the same time. For example, alloying of the catalyst metal by a practically inactive metal (*e.g.* Pt-Au, Pd-Au, Ni-Cu) has been found to increase the hydrogenation activity, whilst the activity for hydrogenolysis decreases[17]. The use of various ruthenium-transition metal alloys seems promising for the partial hydrogenation of benzenes[18].

References

1. P. Sabatier and J.B. Senderens, *C.R. Acad. Sci.*, **124**, 1358 (1897).
2. M. Raney, U.S. Patents 1,563,687 (1927), 1,628,190 (1927), and 1,915,473 (1933); R. Mozingo, *Org. Synth.*, Coll. Vol. 3, 181 (1955); A.A Pavlic and H. Adkins, *J. Am. Chem. Soc.*, **68**, 1471 (1946); H. Adkins and A.A. Pavlic, *Ibid.*, **69**, 3039 (1947); H. Adkins and H.R. Billica, *Ibid.*, 70

695 (1948); X.A. Dominguez, I.C. Lopez, and R. Franco, *J. Org. Chem.*, **26**, 1625 (1961).

3. H.C. Brown and C.A. Brown, *J. Am. Chem. Soc.*, **85**, 1003 (1963); C.A. Brown and H.C. Brown, *Ibid.*, **85**, 1005 (1963).

4. W. Reeve and W.M. Eareckson, *J. Am. Chem. Soc.*, **72**, 3299 (1950).

5. A. Descamps, *C.R. Acad. Sci.*, **67**, 330 (1868); B. de Vries, *K. Ned. Akad. Wet. Proc. Ser.*, **B63**, 443 (1960); J. Kwiatek, *Catal. Rev.*, **1**, 37 (1967); J. Basters, H. van Bekkum, and L.L. van Reijen, *Rec. Trav. Chim. Pays-Bas*, **89**, 491 (1970); J. Basters, C.J. Groenenboom, H. van Bekkum, and L.L. van Reijen, *Ibid.*, **92**, 219 (1973); A. Bergman, R. Karlsson, and R. Larsson, *J. Catal.*, **38**, 418 (1975).

6. E.L. Muetterties and F.J. Hirsekorn, *J. Am. Chem. Soc.*, **96**, 4063 (1974); F.J. Hirsekorn, M.C. Rakowski, and E.L. Muetterties, *Ibid.*, **97**, 237 (1975); E.L. Muetterties, M.C. Rakowski, F.J. Hirsekorn, W.D. Larson, V.J. Basus, and F.A.L. Anet, *Ibid.*, **97**, 1266 (1975).

7. J. Sauer and H. Adkins, *Ibid.*, **59**, 1 (1937).

8. *Cf.* R.L. Augustine, *Reduction*, Marcel Dekker, New York, 1968.

9. K.M.K. Muttzall, *High-pressure hydrogenation of fatty acid esters to fatty alcohols*, Ph. D. Thesis, Delft University of Technology, Delft, 1966.

10. R. Adams, V. Voorhees, and R.L. Shriner, *Org. Synth.*, Coll. Vol. **1**, 463 (1948).

11. H. van Bekkum, J. van Gogh, and G. van Minnen-Pathuis, *J. Catal.*, **7**, 292 (1967); F. van Rantwijk, C.J. Groenenboom, and H. van Bekkum, to be published.

12. E.N. Marvell and T. Li, *Synthesis*, **1973**, 457.

13. J.A. Osborn, F.H. Jardine, J.F. Young, and G. Wilkinson, *J. Chem. Soc. A*, **1966**, 1711.

14. H.S. Broadbent, G.C. Campbell, W.J. Bartley, and J.H. Johnson, *J. Org. Chem.*, **24**, 1847 (1959).

15. G.C. Bond, P.A. Sermon, G. Webb, D.A. Buchanan, and P.B. Wells, *J. Chem. Soc., Chem. Commun.*, **1973**, 444; G.C. Bond and P.A. Sermon, *Gold Bull.*, **6**, 102 (1973).

16. O. Weisser and S. Landa, *Sulphide catalysts, their properties and applications*, Pergamon Press, Oxford, 1973, p 182 ff.

17. J.K.A. Clarke, *Chem. Rev.*, **75**, 291 (1975); V. Ponec, *Catal. Rev.*, **11**, 41 (1975).

18. G.P. Nowack and M.M. Johnson, U.S. Patent, 3,912,787 (1975), *Chem. Abstr.*, **84**, 16861b (1976).

6. Reaction conditions

In most instances, the reaction is carried out in the liquid phase with methanol, ethanol, acetic acid, ethyl acetate or a hydrocarbon as the solvent. The solvent has to be of a high quality in order to avoid poisoning of the catalyst. The use of platinum or palladium oxide as the catalyst requires the presence of a water-miscible solvent, otherwise the water formed upon reduction of the oxide causes agglomeration of the metallic particles. Hydrogenolysis reactions are preferentially carried out in protic solvents, in particular when hetero atoms are involved in the reaction. Aprotic solvents are favourable for the hydrogenation of aromatic and alkenic compounds.

In general, hydrogenations are carried out in neutral media, whereas for hydrogenolysis both acidic and basic media are often used. Furthermore, the hydrogenolysis reaction may be influenced by the addition of acid or base. For instance, the hydrogenolysis of the carbon-halogen bond is facilitated by base, whereas C-O, C-N and C-C hydrogenolysis is enhanced by traces of acid. On the other hand, the rate of hydrogenation is not affected to a great extent by acidic or basic additives. The solvent and the presence of acid or base may drastically influence the selectivity as well as the stereochemistry of the reaction, as demonstrated for various hydrogenations (*e.g.* C=C, C=O) and hydrogenolyses (*e.g.* C-C, C-O, C-N).

PROMOTION AND POISONING[1,2]

The activity of heterogeneous (metallic) catalysts may be influenced by the addition of various reagents. It is possible to make the following subdivision:

i. Interference in the reaction. The above-mentioned promotion or suppression of hydrogenolysis by acid or base may be considered as an example of such interference. In addition, the use of iron(III) or tin(II) salts in the hydrogenation of aldehydes avoids poisoning of the catalyst.

ii. Modification of the catalyst. Partial poisoning of the catalyst has been applied in some cases in order to suppress undesirable parallel or consecutive reactions. This phenomenon has been explained by reference to the fact that the catalyst surface contains various active sites, having quite different properties with respect to activity and selectivity.

It may be noted that in many cases the subdivisions (i) and (ii) are not quite distinct: the co-reagent added will often interfere directly in the reaction and modify the catalyst at the same time. Finally, the support may also play a role, in combination with the reagent added, in promotion and poisoning phenomena.

PRESSURE AND TEMPERATURE

A large number of reactions can be carried out under mild conditions, mostly at room temperature and atmospheric pressure, in particular with platinum, palladium and rhodium as catalysts. Raney nickel catalysts sometimes require a higher temperature and hydrogen pressure. Ruthenium has to be used under high hydrogen pressure (up to 150 atm) in order to get sufficient adsorption of hydrogen on the metal surface. The other heterogeneous catalysts generally require rather drastic reaction conditions. Homogeneous catalysts must be applied under mild conditions, otherwise decomposition of the transition metal complex occurs.

References

1. C. Kemball, in *Catalysis; Progress in Research*, F. Basolo and R.L. Burwell, Jr., Eds., Plenum Press, London, 1973, p 85.
2. M. Freifelder, *Ann. N.Y. Acad. Sci.*, **145**, 5 (1967).

7. Apparatus, procedures, and safety precautions

Reactions at atmospheric pressure may be simply accomplished in a reaction vessel with magnetic stirrer, injection septum, gas-inlet and -outlet tubes, and a revolving tubular device for adding solids. The gas-inlet is connected to a hydrogen burette containing water or paraffin oil as the displacing liquid. The

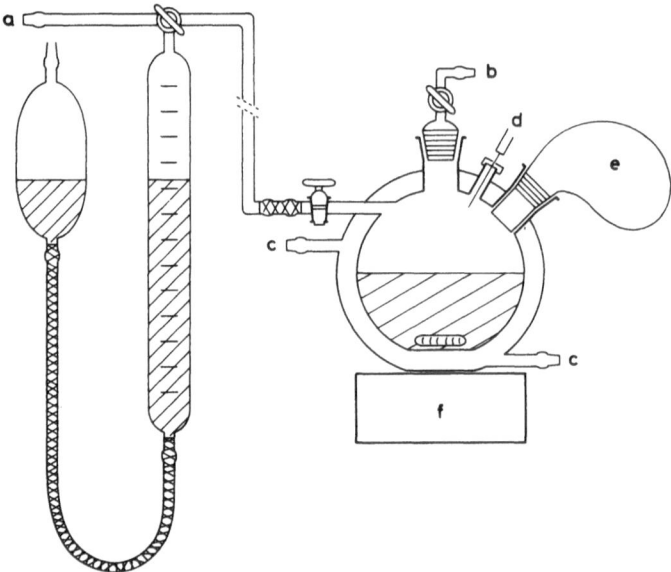

Apparatus for hydrogenations at atmospheric pressure; a. hydrogen; b. vacuum c. thermostate; d. septum; e. revolving device; f. magnetic stirrer.

gas-outlet is connected to a vacuum line. The use of an efficient (magnetic stirrer or mechanical agitation by shaking is important to dissolve the hydrogen into the solution at a sufficient rate and to transport the hydrogen and reactant to the catalyst. Firstly, the catalyst and solvent are placed in the reaction vessel, the system is evacuated and is subsequently filled with hydrogen (without stirring). This procedure is repeated twice in order to remove any oxygen (sometimes it is necessary to purge the system with nitrogen before this procedure, *e.g.* in the case of air-sensitive catalysts). The stirrer is switched on in order to saturate the catalyst and solvent with hydrogen. After the uptake of hydrogen has ceased, the reactant is added (via a syringe or the revolving tubular device) and the reaction is followed by measuring the uptake of hydrogen. The reaction is stopped by switching off the stirrer, evacuating, and purging with nitrogen (or air). After the catalyst has been removed, the product is obtained by the usual work-up procedures. If an accurate measure of hydrogen-uptake is of less importance, *e.g.* in the case of reactants with one reducible function, it may be easier to place reactant, catalyst, and solvent in the reaction vessel at the same time and to work up the reaction mixture after hydrogen-uptake has ceased.

If automatic detection of the uptake of hydrogen is desired, the commercially available Engelhard apparatus (deliverable for various pressure ranges) may be used or the simple automatic hydrogenation apparatus as described in the literature[1]. An alternative to the use of hydrogen gas is the Brown hydrogenator[2] (commercially available) which generates hydrogen *in situ* by the reaction of acid on sodium borohydride.

High pressure hydrogenation and hydrogenolysis may be carried out in a Parr apparatus (up to 3 atm) or in an autoclave (up to 200-300 atm)[3]. Here, the catalyst, solvent and reactant are placed into the apparatus at the same time. Consequently measurement of the uptake of hydrogen by the reactant is less accurate because of the unknown hydrogen uptake of both catalyst and solvent.

For the hydrogenation of very small amounts of material, in determinations of unsaturation, various micro-hydrogenators have been developed[4]. These might also be useful in preliminary experiments for the determination of optimum reaction conditions.

In addition to the usual safety precautions to be taken when working with hydrogen gas, special attention should be paid to the following points:

i. any contact of a catalyst with a mixture of hydrogen and oxygen will lead to ignition;

ii. after the reaction the catalyst will have become pyrophoric because of chemisorbed hydrogen (it may be noted that Raney nickel must be handled under a solvent *at all times* as it is highly pyrophoric).

Consequently, filtration of the reaction mixture, to remove a heterogeneous catalyst, is best carried out without suction. The filter must always be covered with solvent. Before storage or disposal of the recovered catalyst, it must be deactivated by addition of water in order to prevent ignition.

References

1. G.W.H.A. Mansveld, A.P.G. Kieboom, W.Th.M. de Groot, and H. van Bekkum, *Anal. Chem.,* **42**, 813 (1970).
2. C.A. Brown and H.C. Brown, *J. Am. Chem. Soc.,* **84**, 2829 (1962).
3. *Cf.* A.I. Vogel, *Practical Organic Chemistry,* 3rd Ed., Longmans, 1972, pp 866-870.
4. S. Siggia, *Quantitative Organic Analysis,* 3rd Ed., Wiley, New York, 1963, pp 318-341.

II. Hydrogenation

1. Introduction[1]

Hydrogenation is defined as the reductive transformation of a π-bonded system according to:

$$A = B \quad \rightarrow \quad HA - BH$$

in which A and B represent carbon, oxygen or nitrogen. Concerted addition of hydrogen to a π-bond would constitute a $_\sigma 2 + _\pi 2$ process, which is disallowed as far as thermally induced suprafacial reaction is concerned[2]. For that reason, the hydrogenation reaction must proceed in a stepwise manner. 'Ionic hydrogenation' is one possibility[3], but that subject falls outside the scope of the present work. In catalytic hydrogenation processes, the addition of hydrogen also proceeds stepwise, but the intermediates are stabilized by interaction with a catalyst, usually a group VIII metal[4].

As a general rule the reactants — unsaturated compound and hydrogen — have to be brought together at the active site. The first reaction step therefore comprises chemisorption (or, with homogeneous catalysts, coordination) of hydrogen and the π-system.

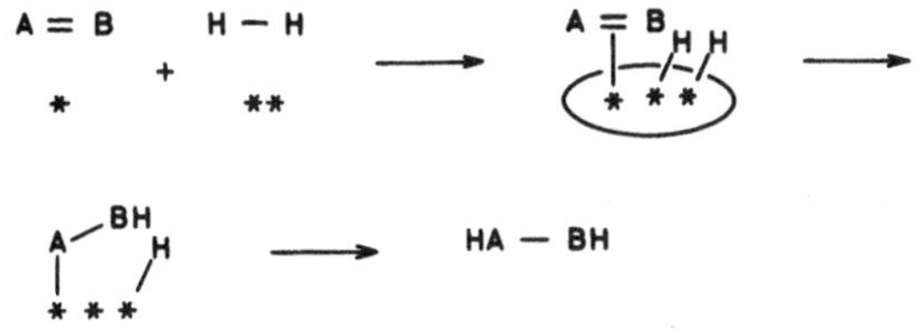

With heterogeneous catalysts, an active site consists of a cluster of metal atoms; a number of metal atoms may therefore be involved in the chemisorption and subsequent hydrogen transfer (sorptive insertion) steps. With homogeneous catalysts, only a single metal atom is available for coordination of the π-system and one or two hydrogen atoms. For that reason, catalytically active complexes are generally coordinatively unsaturated; at least one labile (easily substituted) ligand is usually also present.

Adsorption of hydrogen at a metal surface is usually regarded as a dissociative process (homolytic cleavage of hydrogen), with the formation of atomically chemisorbed hydrogen[5]. Activation of hydrogen by a homogeneous catalyst may involve either heterolytic cleavage (i), homolytic cleavage (ii), or oxidative addition (iii) depending on the character of the catalyst[6].

$$Ru^{III}Cl_6{}^{3-} + H_2 \rightleftharpoons Ru^{III}HCl_5{}^{3-} + H^+ + Cl^- \qquad \text{(i)}$$

$$2\, Co^{II}\ (CN)_5{}^{3-} + H_2 \rightleftharpoons 2\, Co^{III}H(CN)_5{}^{3-} \qquad \text{(ii)}$$

$$Rh^{I}Cl(PPh_3)_3 + H_2 \rightleftharpoons Rh^{III}ClH_2(PPh_3)_3 \qquad \text{(iii)}$$

Chemisorption of a π-system is generally thought of as being brought about by overlap of the p-orbitals of the π-bond and the spd-orbitals of the metal atom(s). Two models have been advanced for the coordination of alkenes. In one the alkene-metal bond is represented by σ-type overlap between the π-orbitals on one hand and the metal orbitals on the other (**I**), in the other it is represented by a metallocyclopropane ring (**II**)[7]. From the available data we may conclude that coordination involves weakening of the C-C π-bond and rehybridization of the alkenic carbon atoms[7]. Backbonding (overlap of the occupied metal orbitals and the empty alkene π^* orbital) also contributes to the weakening of the alkenic bond[8]. It seems that model **II** is only realistic for alkenes which exhibit extraordinarily strong backbonding[7], such as tetracyanoethene. With heterogeneous catalysts, models **I** and **II** are also used as well as the 1,2-diadsorbed structure **III**.

I   II   III

Chemisorption involves an equilibrium between free and coordinated molecules. With homogeneous catalysts, this process should be treated as a displacement equilibrium. Chemisorption on a heterogeneous catalyst can be described by the Langmuir adsorption isotherm[9]:

$$\Theta_{AB} = \frac{b_{AB}[\,AB\,]}{1 + b_{AB}[\,AB\,] + \Sigma bc}$$

in which Θ_{AB} is the fraction of active catalyst covered with AB, b_{AB} the adsorption constant and Σbc is the sum of the contributions of the solvent, the hydrogenated product, and the hydrogen to the denominator of the Langmuir expression.

It should be noted that the effect of hydrogen on the adsorption of reactant will be disregarded. This is generally correct since only sites which can accommodate substrate as well as hydrogen are active.

If two unsaturated compounds A=B and C=D compete for the catalyst surface, the Langmuir expression becomes:

$$\Theta_{AB} = \frac{b_{AB}[AB]}{1 + b_{AB}[AB] + b_{CD}[CD] + \Sigma bc}$$

or, if $b_{AB}[AB] + b_{CD}[CD] \gg 1 + \Sigma bc$, i.e. AB and CD occupy the active surface almost completely,

$$\Theta_{AB} = \frac{b_{AB}[AB]}{b_{AB}[AB] + b_{CD}[CD]} = \frac{K[AB]}{K[AB] + [CD]}$$

in which $K = b_{AB}/b_{CD}$ (the adsorption equilibrium constant).

MECHANISM

Mechanistic studies of hydrogenation reactions reveal that the transfer of the two hydrogen atoms proceeds stepwise; transfer of the first hydrogen atom is almost without exception reversible. The half-hydrogenated state IV therefore plays a crucial role in homogeneously and heterogeneously catalyzed hydrogenation reactions[6,10].

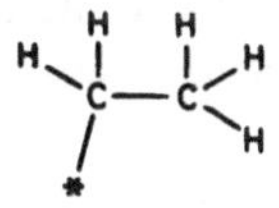

IV

KINETICS

If no real competition between hydrogen and the reactant occurs on the catalytic surface, the rate equation is

$$-\frac{d[AB]}{dt} = r_{AB}' = k_{AB}' \Theta_{AB},$$

in which r_{AB}' and k_{AB}' are the reaction rate and pseudo reaction rate constan per weight amount of catalyst.

This gives upon combination with the Langmuir adsorption isotherm:

$$r_{AB}' = \frac{b_{AB}\,[AB]\,k_{AB}'}{1 + b_{AB}\,[AB] + \Sigma bc}$$

appropriately called the Langmuir rate equation.

No general treatment for homogeneous hydrogenation can be given, owin to the widely diverging reaction schemes[10].

SELECTIVITY

When two unsaturated compounds A=B and C=D are hydrogenated in competi tion, the selectivity can be calculated from the Langmuir rate equations:

$$r_{AB}/r_{CD} = \frac{\Theta_{AB}k_{AB}'}{\Theta_{CD}k_{CD}'} = K\,\frac{[AB]\,k_{AB}'}{[CD]\,k_{CD}'}$$

Thus, the position of the chemisorption equilibrium and the relative rates o: hydrogen transfer contribute equally to the selectivity. If the adsorptior equilibrium constant is extremely large, A=B will monopolize the catalys surface. In such a case the reaction of C=D does not occur until all A=B has beer consumed.

References

1. See also: Commission on Colloid and Surface Chemistry, Definitions terminology and symbols in colloid and surface chemistry, Part II, hetero geneous catalysis, International Union of Pure and Applied Chemistry Oxford, 1974.
2. *Cf.* R.B. Woodward and R. Hoffmann, *The conservation of orbita symmetry*, Verlag Chemie, Weinheim, 1970.
3. See: D.N. Kursanov, Z.N. Parnes and N.M. Loim, *Synthesis*, 1974, 633.
4. Some microorganisms also catalyze alkene hydrogenation; see: H. Simon B. Rambeck, H. Hashimoto, H. Günter, G. Nohynek, and H. Neumann *Angew. Chem.*, **86**, 675 (1974); B. Rambeck and H. Simon, *Ibid.* B. Rambeck, Thesis, Technische Universität, München, 1974.
5. *Cf.* P.C. Aben, H. van der Eijk, and J.M. Oelderik, *Proc. Int. Congr. Catal. 5th, 1972*, **1**, 717 (1973).
6. J. Halpern, *Adv. Chem. Ser.*, **70**, 1 (1968); M.M.T. Khan and A.E. Martell *Homogeneous catalysis by metal complexes*, Vol. I, Academic Press, New York, 1974.

7. See: R.D.W. Kemmitt in *MTP Int. Rev. Sci., Inorg. Chem. Ser. One,* **6,** Trans. Met. Pt. 2, M.J. Mays, Ed., Butterworths, London, 1972, p 227, and references cited therein.
8. A. van der Ent, Thesis, University of Nijmegen, The Netherlands, 1973.
9. I. Langmuir, *Trans. Faraday Soc.,* **17,** 607 (1922).
10. For further details regarding the mechanisms of homogeneously catalyzed hydrogenation; see B.R. James, *Homogeneous hydrogenation,* Wiley-Interscience, London, 1973.

2. Hydrogenation of alkenic double bond

Saturation of the alkenic double bond can be effected in the presence of many catalysts, usually under ambient conditions. Historically, palladium seems to be the preferred catalyst, although the tendency of this metal to effect double bond migration makes it less suitable in many cases. Platinum, rhodium and Raney nickel also find extensive application; ruthenium and some homogeneous catalysts are to be preferred in special cases. As a general rule, the catalytic activity decreases in the order Pt > Pd > Rh > Ru ≫ Ni for equal surface of metal.

STRUCTURE AND REACTIVITY

As has been explained previously, the rate of conversion depends on chemisorption and hydrogen transfer. In simple heterogeneously catalyzed hydrogenations zero order reaction kinetics are usually observed, indicating occupation of all suitable sites by alkene. Although chemisorption does not seem a rate-limiting factor in heterogeneous hydrogenation, homogeneous hydrogenation involves displacement of a ligand from the catalyst[1]. As a result, first or broken order reaction kinetics are observed.

Although electronic factors have considerable influence on chemisorption and coordination, the effect on the rate of hydrogen transfer appears to be small[2]. In many cases conjugated double bonds are, in homogeneous hydrogenation, more reactive than isolated ones[1] (the phenomenon is, however, far from general[1,3]). The enhanced polarizability of the π electron system and the availability of low-lying antibonding orbitals might be important factors in decreasing the free energy of activation of the displacement process.

No reliable data regarding the influence of steric factors on chemisorption are available. Comparable transition metal alkene complexes have been studied, however[4]. Complexes of terminal alkenes are more stable than complexes of internal alkenes, and coordination is known to be stabilized by ring strain in the free ligand[5]. The influence of steric factors on the rate of hydrogenation is

34

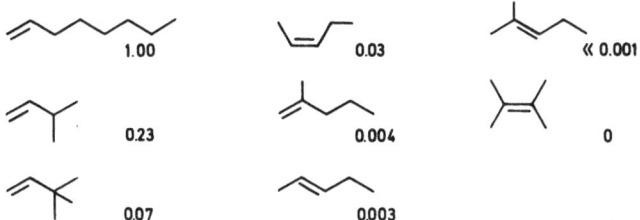

Relative rates of hydrogenation over P-2 nickel in ethanol at 25° and 1 atm[6].

considerable. Reactivity decreases with increasing substitution of the double bond: terminal > *cis*-internal > *trans*-internal > trisubstituted > tetrasubstituted alkene, unless considerable strain is released in the course of the reaction. The heterogeneous P-2 nickel catalyst (*cf*. figures) is reputed to be extremely sensitive towards structural influences[6]. It is interesting to note that a study[7] of alkene hydrogenation kinetics with chlorotris(triphenylphosphine)-

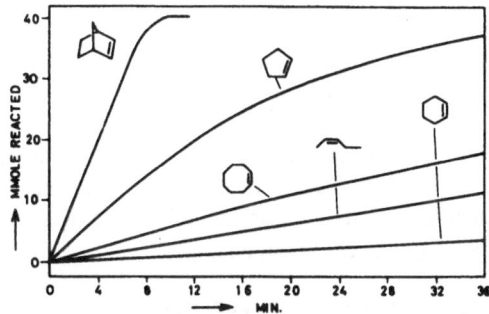

Hydrogenation of disubstituted alkenes over P-2 nickel at 25°: 40.0 mmole of substrate and 5.0 mmole of catalyst (1.25 mmole for bicyclo [2.2.1] hept-2-ene and cyclopentene)[6].

rhodium(I) as the homogeneous catalyst revealed a decrease of $\Delta H^{\neq}$ with increasing substitution, reflecting a release of strain in the transition state. This effect was overruled, however, by a decrease of $\Delta S^{\neq}$ due to the rigidity of the transition state for highly substituted alkenes. Some homogeneous hydrogenation catalysts exhibit a marked preference for hydrogenation of terminal double bonds. Chlorohydridotris(triphenylphosphine)ruthenium(II) constitutes the best-known example: the catalyst is nearly inactive for the hydrogenation of internal double bonds[8]. Since isomerization and exchange of internal alkenes still occur, the effect should originate in the transfer of the second hydrogen atom and has been attributed to steric hindrance[8]. Some other homogeneous catalysts exhibit similar behaviour in a less pronounced way[3,9,10].

Homogeneous hydrogenation of tetrasubstituted alkenes proceeds extremely sluggishly, if at all. Application has been found in the preparation of bicyclic tetrasubstituted alkenes such as 1 and 2[11].

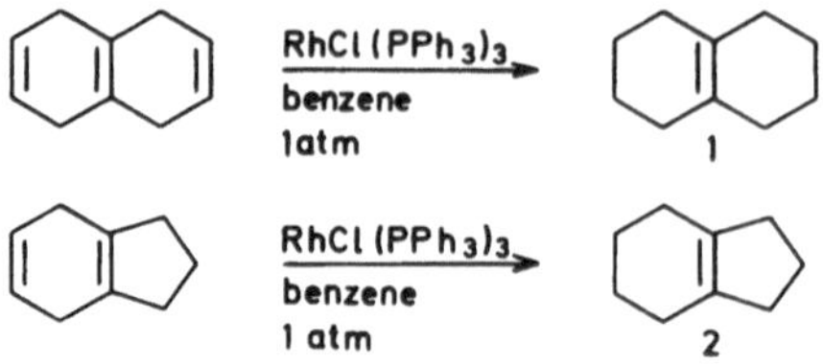

Release of ring strain can also be an important factor in determining the hydrogenation rate, as shown by the hydrogenation of cis-alkenes over P-2 nickel in ethanol; relative rates of hydrogenation: norbornene 1.0, cyclopen tene 0.26, (Z)-2-pentene<0.01[6].

STEREOCHEMISTRY

Suprafacial (cis) addition of hydrogen would be expected, since transfer of both hydrogen atoms from the catalyst surface to the reactant is usually assumed[12]. This is nearly always the case with palladium, rhodium, ruthenium or nickel as the catalyst. In hydrogenation over platinum, however, antara-facial (trans) addition of hydrogen sometimes contributes[13-16], the reaction mechanism is still under discussion. The hydrogenation of dimethyl bicyclo-[2.2.2]oct-2-ene-2,3-dicarboxylate (3) constitutes an example[16].

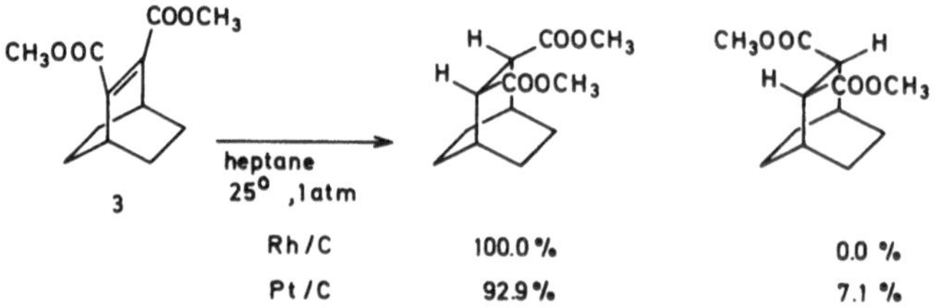

Rh/C	100.0 %	0.0 %
Pt/C	92.9 %	7.1 %

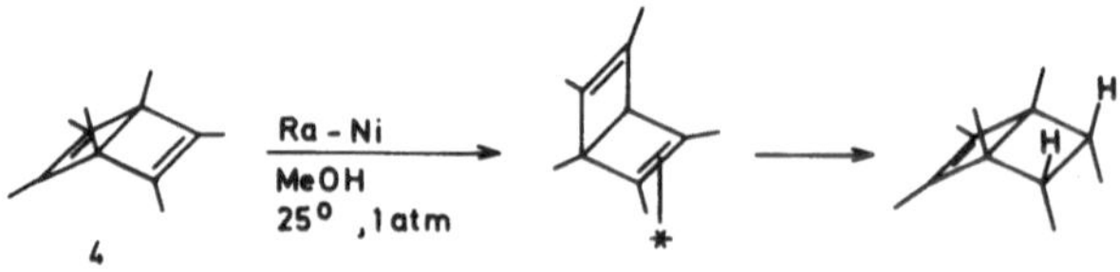

If two topologically different ways of suprafacial addition exist, the reaction pathways should be regarded as involving a chemisorption equilibrium. The ratio in which the diastereomeric products are formed reflects the position of the chemisorption equilibrium as well as the relative rates of hydrogen transfer, which are both usually decided by steric factors. Thus, in small bicyclic com-pounds exo addition is usually favoured over endo addition[17], although the latter would lead to a less strained product, as, for example, the hydrogenation of hexamethylbicyclo[2.2.0]hexa-2,5-diene (4) over Raney nickel[13]. In steroid compounds, the angular methyl groups at the 10- and 13-positions exclude contact between the β-side and the catalyst surface. Accordingly, α-hydrogenation of alkenic steroids is nearly always observed. 3-Methoxyestra-1,3,5(10),8(14)-tetraene-17β-ol[18] (5) and methyl 3-ketoetiochola-4,9(11),16-trienoate[19] (6) are examples.

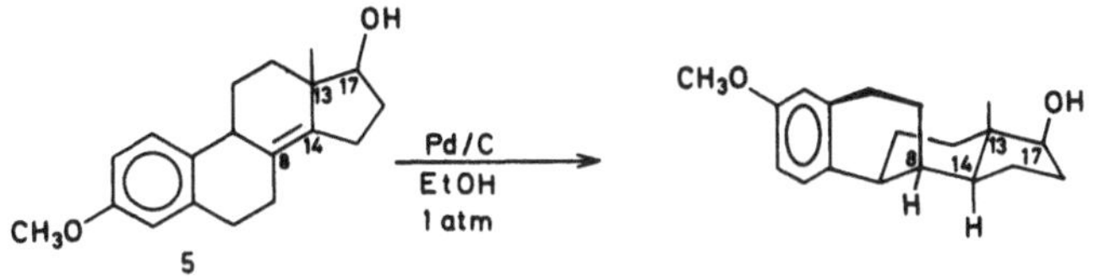

α-Hydrogenation plays an important role in many sythetic approaches to the steroid skeleton[20]. In the total synthesis according to Velluz[21], *trans*-fusion of the C and D rings was effected by means of stereospecific α-hydrogenation of the bicyclic fragment 7.

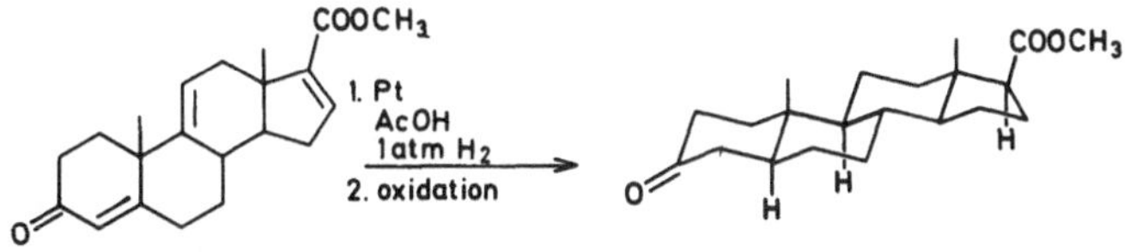

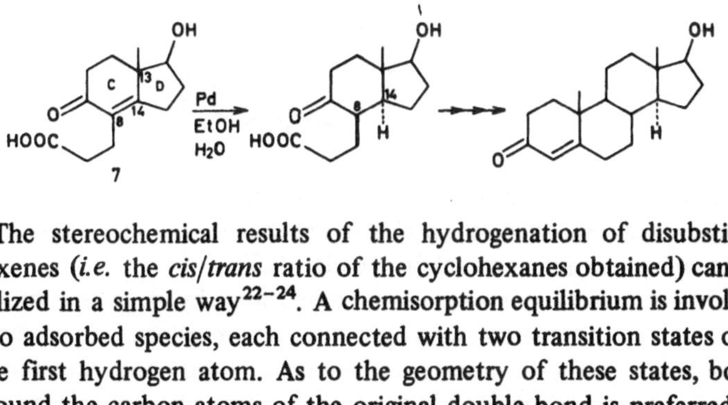

The stereochemical results of the hydrogenation of disubstituted cyclohexenes (*i.e.* the *cis/trans* ratio of the cyclohexanes obtained) cannot be rationalized in a simple way[22-24]. A chemisorption equilibrium is involved between two adsorbed species, each connected with two transition states of transfer of the first hydrogen atom. As to the geometry of these states, bond eclipsing around the carbon atoms of the original double bond is preferred. As a result the ring is in a boat conformation[22,24]. This model predicts satisfactorily several stereochemical trends observed in the hydrogenation of 1,4- and 2,3-substituted cyclohexenes[22,25,26] but should not be regarded as conclusive[26]. The stereochemical trend may become rather dependent on the reaction temperature, when bulky substituents are present as in, for example, 4-methyl-1-*t*-butylcyclohexene (8)[26,27].

Steric control is also important in homogeneous hydrogenation. Deuteration of 2,5-dimethoxy-2,5-dihydrofuran (9) with chlorotris(triphenylphosphine)-

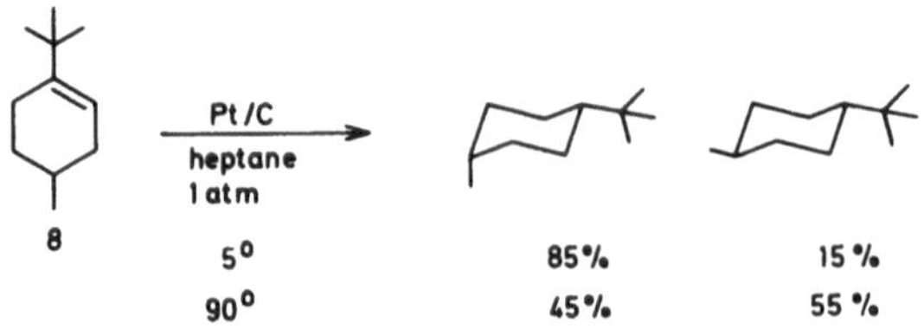

	85%	15%
5°		
90°	45%	55%

rhodium(I) resulted in suprafacial addition of deuterium at the least hindered face[28]. It should be noted that the Wilkinson catalyst is generally very useful for clean deuteration of alkenes[29].

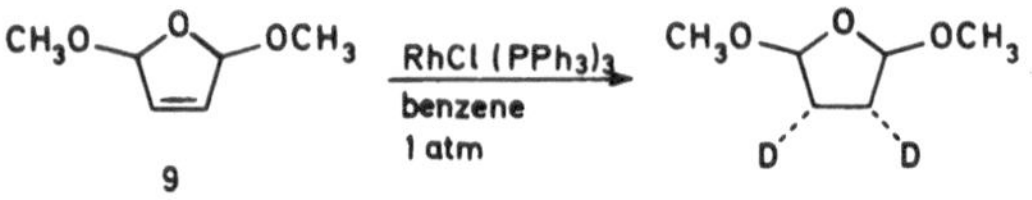

Bond formation between the catalyst and a polar functional group present in the substrate may also influence the steric course of the hydrogenation ('haptophilic control'[30]). The stereospecific hydrogenation of 1-hydroxy methyl-6-methoxytricyclo[8.4.0.0^{4,9}]tetradeca-4,6,8,10-tetraene alkali metal alcoholate (10) with Wilkinson catalyst is an example[31] (the alcohol did not

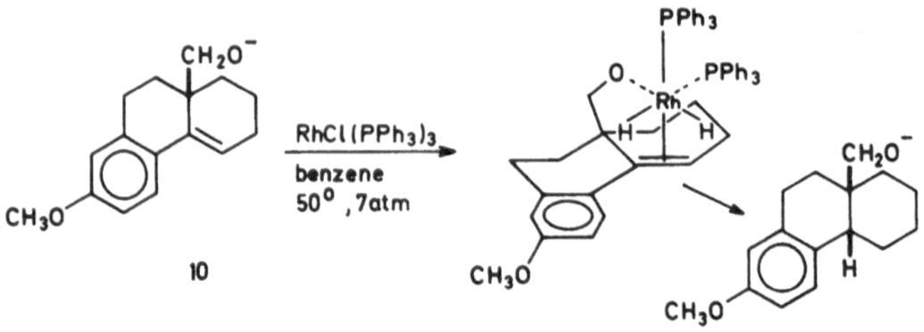

react). Enantioselective hydrogenation in the presence of a chiral catalytic system is a very special case. Success is probably dependent on a subtle interplay between steric and haptophilic control (see section II.7).

The stereochemical picture is sometimes complicated by migration of the double bond, or by alternative hydrogen transfer reactions. Antarafacial addition of hydrogen in the presence of platinum has already been mentioned. A discussion of the various 'topside addition' mechanisms which have been put forward[32] falls outside the scope of the present work.

With palladium as the catalyst, participation of the solvent is sometime observed: hydrogenation of dimethyl bicyclo[2.2.2]oct-2-ene-2,3-dicarboxylate (3) in the presence of a trace of strong acid resulted in up to 57% antarafacial addition. The effect is exclusive to palladium[15b,16]. Hydrogenation of 4-(3,5 dimethyl-4-isoxazolylmethyl)-7,7a-dihydro-1β-hydroxy-7a, β-methyl-5 (6H) indanone (11) over palladium in the presence of perchloric acid afforded mainly the antarafacially hydrogenated product[33], also indicating solvent parti

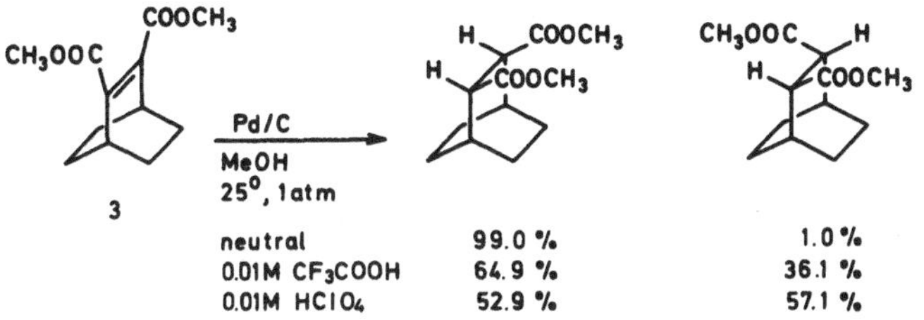

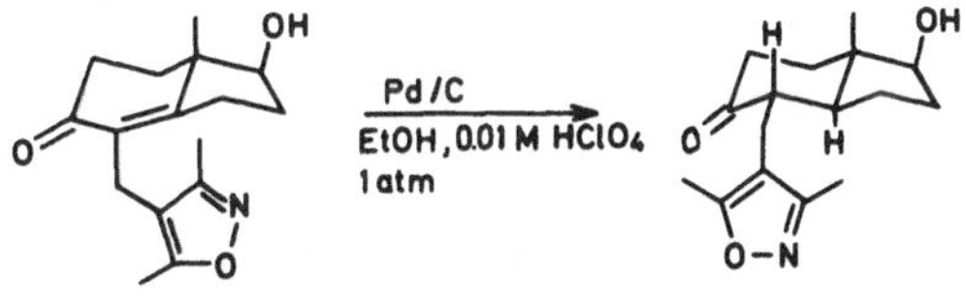

11

cipation. The influence of acid on the stereospecificity of palladium catalyzed hydrogenations is ascribed to the attack of a proton from the solution on the intermediately formed alkyl surface complex[15]. The presence of a polarizable

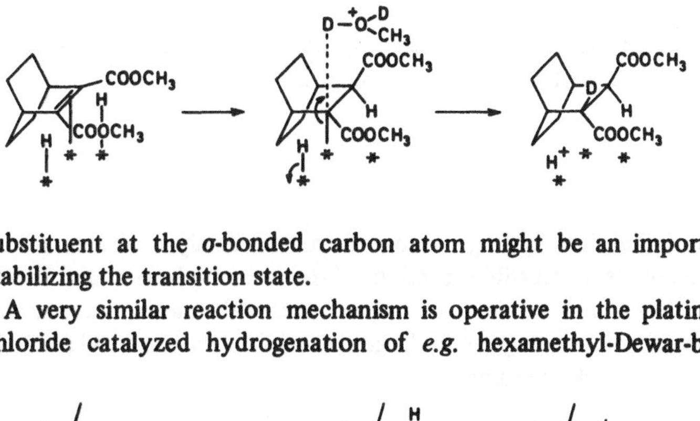

substituent at the σ-bonded carbon atom might be an important factor in stabilizing the transition state.

A very similar reaction mechanism is operative in the platinum(II)-tin(II) chloride catalyzed hydrogenation of *e.g.* hexamethyl-Dewar-benzene (4)[13].

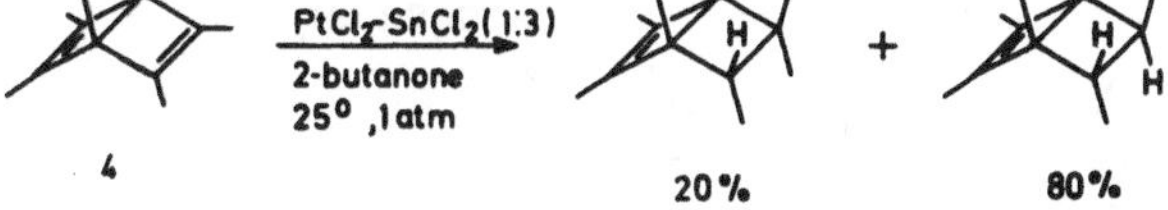

Hydrogenation of 4-*t*-butylmethylenecyclohexane (12) in the presence of the platinum-tin catalyst gave 95% of *trans*-4-*t*-butylmethylcyclohexane[34]. Coordination of the exocyclic double bond from the equatorial direction (favoured for steric reasons) followed by suprafacial addition of hydrogen would lead to the *cis*-isomer (axial methyl group). Since 4-*t*-butyl-1-methylcyclohexene seems not to be involved[34,35], the result indicates either equilibration of the cyclohexylmethyl complexes A and B[34] (probably via decoordination and an equatorially coordinated boat conformer) or protonolysis of a cyclohexyl complex

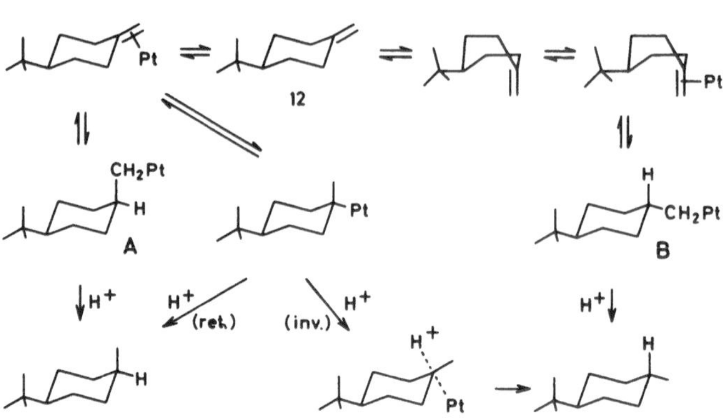

with inversion. Hydrogenation of 3-methylenecholestane **(13)** similarly afforded 3β-methylcholestane[34].

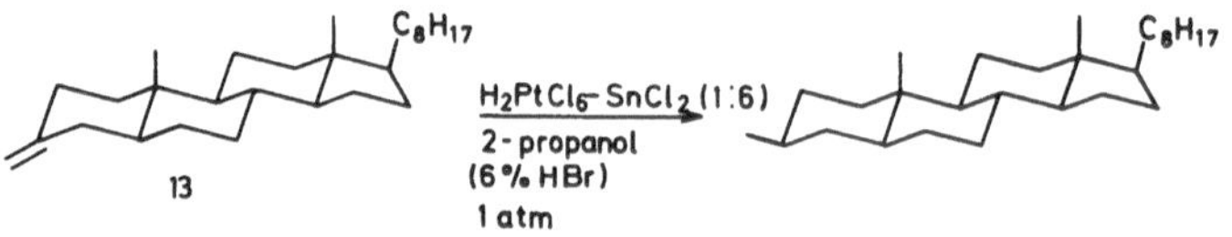

Platinum-tin catalyzed antarafacial hydrogenation of methyl (*E*)-2-phenyl-
ethenecarboxylate (methyl cinnamate) has recently been established[36].

ISOMERIZATION

The transfer of the first hydrogen atom from the catalyst surface to the
chemisorbed alkene is a reversible reaction; *cis-trans* isomerization and double
bond migration result if a different hydrogen atom is abstracted in the reverse
reaction[37]. The palladium catalyzed hydrogenation of 4-*t*-butylmethylenecyclo-
hexane[38] **(12)** is a very clear example.

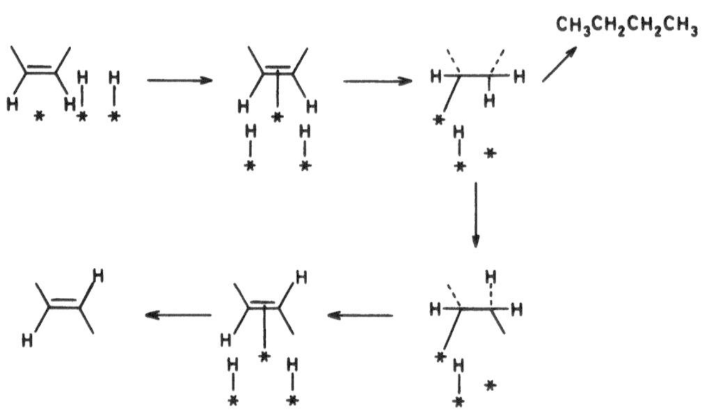

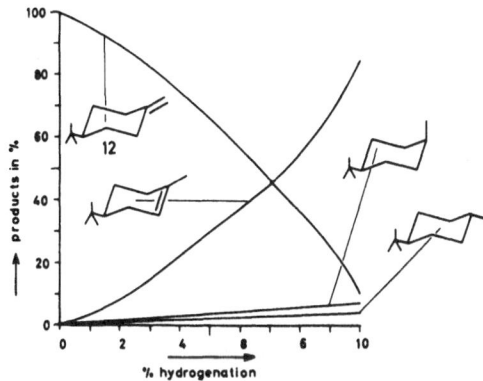

Hydrogenation of 4-t-butylmethylenecyclohexane over Pd/C in ethanol.

The influence of the catalyst metal on the degree of isomerization is considerable. For 1-pentene the order palladium $\gg$ ruthenium $>$ rhodium $>$ platinum
$\geqslant$ iridium has been established[39]. (+)-1-p-Menthene (14) has been used as a
model compound for the investigation of double bond migration[40]. Hydro-

<table>
<tr><td></td><td>cyclohexane
ambient conditions</td><td></td><td></td></tr>
</table>

14

Pd/C	15.3 % $\frac{\alpha}{\alpha_o}$= 0.45 racemization 47 %
Pt/C	28.4 % $\frac{\alpha}{\alpha_o}$= 0.70 racemization ‹ 3 %

Racemization of (+)-1-p-menthene during hydrogenation.

genation of **14** over palladium resulted in extensive migration and concomitant
racemization. On the other hand, hydrogenation over platinum was not accompanied by racemization. Alternative isomerization mechanisms have been put
forward on the basis of deuteration experiments, such as the π-allyl[41] and roll-
over[42] mechanisms. The possibility of a metal assisted suprafacial [1,3] sigma-
tropic shift of hydrogen has been discussed on the basis of deuteration
experiments[43]. Such a process would constitute an example of straightforward
forbidden-to-allowed catalysis.

The degree of isomerization is dependent on the relative rates of abstraction
of a hydrogen atom from the alkyl species and transfer of the second hydrogen
atom. Geometric factors may sometimes play an important role, as became
clear from the results of deuteration of various cycloalkenes in the presence of
Wilkinson catalyst[44]. Whereas cyclohexene gave clean addition of two deuterium
atoms, medium-sized rings like cyclooctene and cyclododecene exhibited
extensive isotopic exchange. Allyl ethers also exhibited isomerization in
the presence of chlorotris(triphenylphosphine)rhodium(I) with formation of

41

the 1-propenyl ethers[45]. Allyl ethers can therefore be used as alcohol protecting groups; their removal involves treatment with Wilkinson catalyst followed by mild acid. Allyl ethers isomerize significantly faster than the corresponding

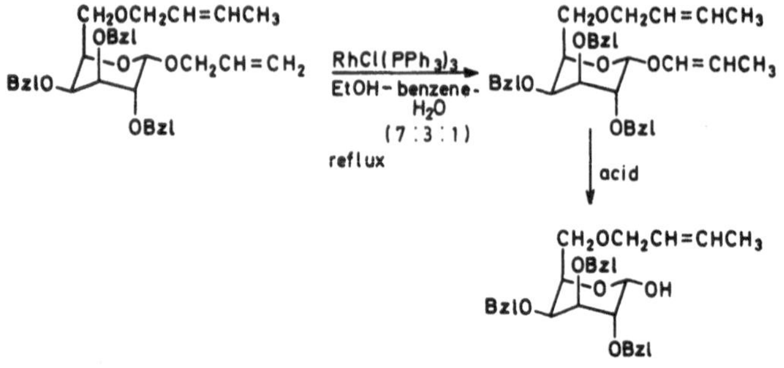

but-2-enyl ethers, thus allowing selective removal of the allyl group[46].

Double bond migration, followed by desorption and renewed adsorption at the opposite face will give rise to overall antarafacial hydrogen addition in cyclic compounds. Thus, hydrogenation of 1,2-dimethylcyclohexene over

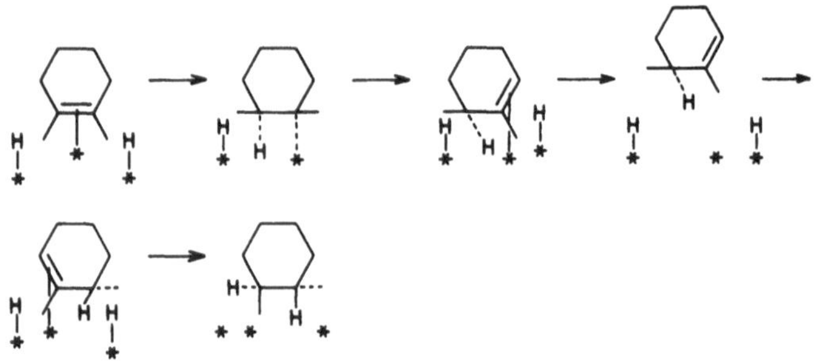

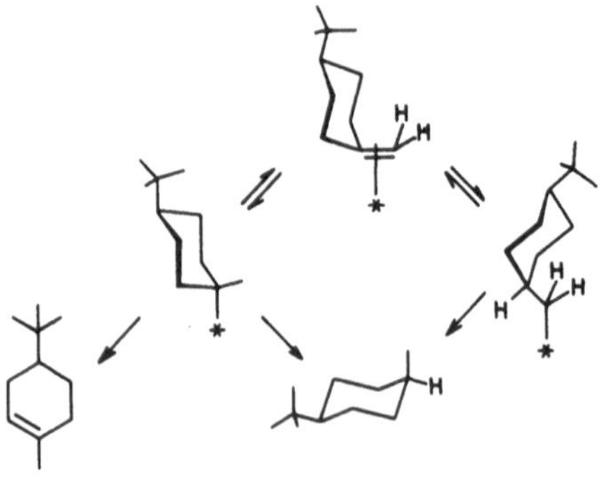

palladium gave mainly *trans*-1,2-dimethylcyclohexane, whilst with platinum as the catalyst the *cis*-isomer was obtained[47]. Hydrogenation of 1,2-dimethyl-cyclopentene revealed a somewhat different pattern: Raney nickel gave preferently *cis*-1,2-dimethylcyclopentane, whereas palladium and platinum gave about 70% of the *trans*-isomer[48]. A similar problem is encountered in the hydrogenation of 4-*t*-butylmethylenecyclohexane for example. Chemisorption or coordination of the exocyclic double bond is for steric reasons expected to occur mainly from the equatorial direction (the *t*-butyl group acts as a 'holding group'), with formation of cis-4-t-butyl-1-methylcyclohexane (methyl group axial). Reaction via the isomerized alkene 4-t-butyl-1-methylcyclohexene, on the other hand, should give rise to a mixture of the isomers (cf. Table III). Hydrogenation of 4-t-butylmethylenecyclohexane with platinum[38], rhodium[34,38] or chlorotris(tri-o-tolylphosphine)rhodium(I)[34] gave mainly the cis-product,

Table III: The hydrogenation of 4-t-butylmethylenecyclohexane[a]

Catalyst	p_{H_2} (atm)	cis/trans
Pt	0.25	6.7
	100	1.6
RhCl(PPh₃)₃	1-100	2.0
Rh(CO)(H)(PPh₃)₂	0.5	0.28
	40	1.8

a: Cf. ref[49] and references cited therein.

while most other catalysts gave mixtures. The ratio of cis and trans products has been used as a mechanistic probe for changing reaction pathways as a function of the conditions; in this way, details of heterogeneous and homogeneous hydrogenation mechanisms have been elucidated[49,75].

POLYENES

Selective partial hydrogenation of polyenes is of considerable preparative and commercial interest. Success depends on the nature of the polyene, as well as on a careful choice of catalyst and conditions.

Selective formation of a monoene from a corresponding diene requires preferential hydrogenation of diene in the presence of monoene ('molecular queueing'), as well as regioselective conversion of one dienoid double bond. The simplest case occurs if, for example for geometric reasons, simultaneous chemisorption of both alkenic bonds is excluded. A chemisorption equilibrium then exists between the two double bonds. The first hydrogenation step may be regarded as a competition reaction; the regioselectivity is determined in the usual way by the position of the chemisorption equilibrium as well as by the relative rates of hydrogen transfer. The final result may depend strongly on the

Ra-Ni	100 %	0 %
Pd /C	85 %	15 %
Ru/C	66 %	34 %
Rh/C	52 %	48 %
Pt /C	47 %	53 %
Os	32 %	68 %

catalyst, as was the case with dimethyl tetramethylbicyclo[2.2.0]hexa-2,5-diene-5,6-dicarboxylate[50] **(15)**. Hydrogenation did not proceed beyond the dihydro stage under ambient conditions and was entirely *cis, exo*. It is sometimes possible to influence the regioselectivity by the introduction of bulky groups. The hydrogenation of di-*t*-butyl tetramethylbicyclo[2.2.0]hexa-2,5-diene-5,6-dicarboxylate is an example[50].

R = COOMe   42%   58%
R = COOtBu   5%   95%

The effect of ring strain on the rate of hydrogenation has already been mentioned; ring strain can also exert an important influence on the regioselectivity. Thus, hydrogenation of 5-methylenenorbornene **(16)** over P-2 nickel resulted in preferential saturation of the strained endocyclic double bond[6], although the exocyclic double bond would seem to be more easily accessible.

With homogeneous catalysts, monodentate chemisorption of polyenes seems to be the rule rather than the exception. Contrary to heterogeneous hydrogenation, only a single metal atom is available, bidentate coordination would therefore require two free coordination sites per complex molecule and, in many cases, an unfavourable conformation of the dienic system. In this way, a possibility for selective partial hydrogenation is created which would sometimes be difficult to accomplish with a heterogeneous catalyst. An example is found in the selective partial hydrogenation of carvone **(17)** in the presence of

chlorotris(triphenylphosphine)rhodium(I)[51]. Bidentate coordination comes to the fore, however, with cyclodienes, in some cases to the extent of causing catalyst poisoning[52].

With heterogeneous catalysts, bidentate chemisorption involving two sites will usually occur if it is geometrically feasible. Since both double bonds are coordinated, the regioselectivity will be determined by the relative rates of the corresponding surface reactions. In the hydrogenation of 2-methylhexa-1,5-

diene (18), 4-vinylcyclohexene[53] (19) and β-ionone over P-1 nickel[54], the least substituted double bond was preferentially attacked.

The use of the dicarbonyl(η^5-cyclopentadienyl)iron moiety as an alkene protecting group is a recent development. The least hindered double bond is preferentially protected. Thus, 4-vinylcyclohexene (19) gave vinylcyclohexane upon hydrogenation, oct-1-ene-4-yne (20) gave 1-octene. The alkenes were liberated by treatment with sodium iodide[55].

Selective hydrogenation of the diene to the monoene stage requires monopolization of the surface by diene. The chemisorption equilibrium should favour the diene, and displacement of monoene from the surface should proceed considerably faster than hydrogen transfer to chemisorbed monoene. The hydrogenation of 1,5,9-cyclododecatriene (21) to cyclododecene in the presence of a homogeneous ruthenium catalyst is an interesting example[56].

Selective hydrogenation of various dienes to the monoenes could be accomplished with a cationic rhodium complex as the catalyst[76]. Selective saturation of one double bond in the triglyceride of (Z,Z,Z)-linolenic acid, yielding a (Z,Z)-linoleate isomer is commercially important. Extensive precautions have to be taken against migration of the skipped double bonds, cis-trans isomerization and overhydrogenation. Since the rates of the surface reactions are of the same order of magnitude, any selectivity has to be achieved through monopolization of the catalyst surface by linolenate[57]. Industrially nickel and copper are applied as catalysts.

45

The problems met in the selective hydrogenation of alkenes bearing unsaturated functional groups are very similar to those encountered in the partial hydrogenation of polyenes.

Because of the high reactivity of the alkenic double bond, selective hydrogenation in the presence of carbonyl groups or aromatic systems usually presents no special difficulty, provided that mild conditions are maintained.

Palladium exhibits low activity for the reduction of aliphatic aldehydes and ketones, and is therefore the preferred catalyst for the selective hydrogenation of alkenic aldehydes and ketones. An example is the hydrogenation of 2-benzylidenecyclopentanone (22), which, with palladium as the catalyst, gives mainly 2-benzylcyclopentanone; with platinum the carbonyl function is also attacked[58]. Selective hydrogenation of crotonaldehyde and cinnamaldehyde

over borohydride reduced palladium chloride has been reported[59]. Partial deactivation of the catalyst by boride might also enhance selectivity.

Homogeneous catalysts have been applied with good results; the selective hydrogenation of 1,4-naphthoquinone leading to 1,2,3,4-tetrahydro-1,4-dioxonaphthalene in the presence of chlorotris(triphenylphosphine)rhodium(I) is a striking example[60]. Attempts have been made to accomplish selective

hydrogenation of alkenic aldehydes in the presence of homogeneous catalysts, but decarbonylation and concomitant deactivation of the catalyst resulted[61]. Hydrogenations of alkenic aldehydes in the presence of iron[62] and rhodium[63] carbonyl complexes are, however, promising techniques.

HYDROGENOLIZABLE GROUPS

Hydrogenation of alkenes bearing hydrogenolizable functions proceeds selectively in most cases. The use of palladium as the catalyst should be avoided if sensitive groups are present. Hydrogenolysis of carbon-oxygen bonds is inhibited by basic or neutral conditions, whilst halides remain preserved in

46

acidic medium under carefully controlled conditions. Allyl and vinyl systems are especially sensitive and require special care. Homogeneous catalysts have also been applied successfully. The low tendency of homogeneous catalysts to effect hydrogenolysis is not unexpected, considering that hydrogenolysis usually requires participation by at least two metal atoms, while homogeneous catalysts have only a single metal atom available.

Rhodium and ruthenium are the preferred catalysts for the hydrogenation of sensitive alkenes such as toxol[64] (23); selective hydrogenation of allyl alcohols,

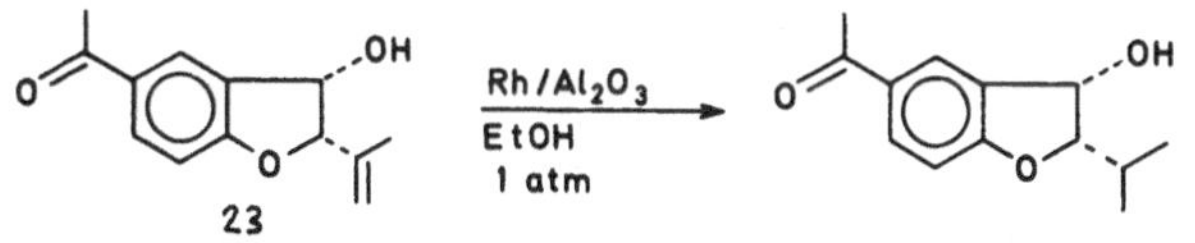

ethers and esters over P-1 nickel has been described[65]. Surprisingly, excellent results have been obtained with borohydride reduced palladium for the hydrogenation of vinyl- and allyloxy compounds[59]. Rhodium on alumina has been recommended as a catalyst for the selective hydrogenation of chloroalkenes; selectivity increases with the distance between the chlorine atom and the double bond[66]. Comparable studies with other catalysts have not, however, been conducted.

Selective hydrogenation of 1,2-bis(trimethylsilyloxy)cycloalkenes over palladium, platinum and rhodium has been claimed[67] as a reaction step in the preparation of cycloalkane-1,2-diols. Confirmation of these results has, however, proved to be difficult[68].

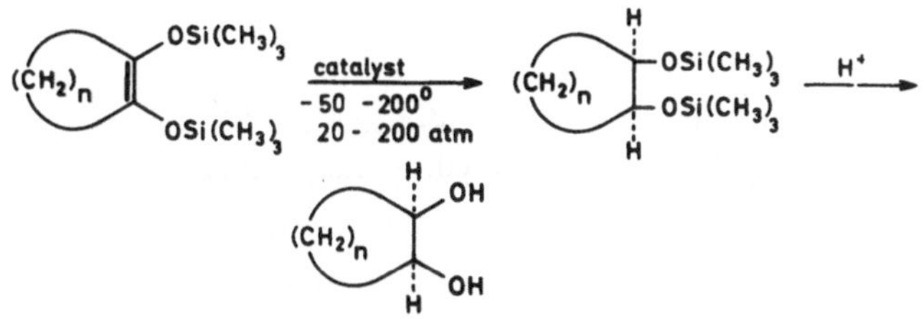

Homogeneous catalysts, especially chlorotris(triphenylphosphine)rhodium(I) have been applied with excellent results. Hydrogenation of β-nitrostyrene and p-nitro-β-nitrostyrene in the presence of Wilkinson catalyst resulted in selective saturation of the alkenic double bond[69].

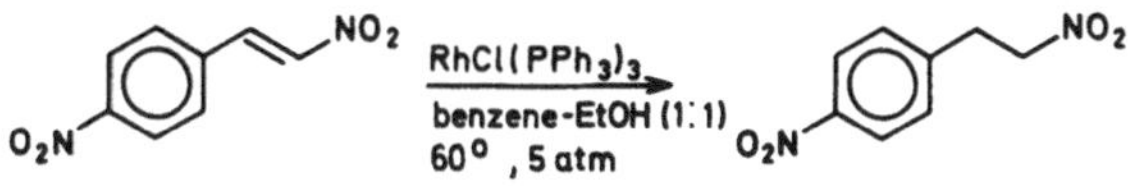

Linalool (24) and testosterone[70] (25) provide further examples of hydrogenation of sensitive compounds in the presence of Wilkinson catalyst[71].

47

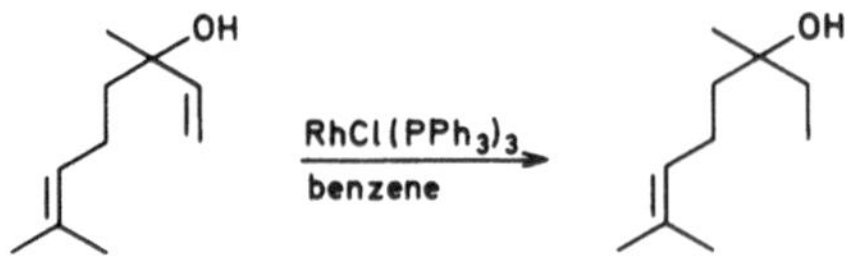

24

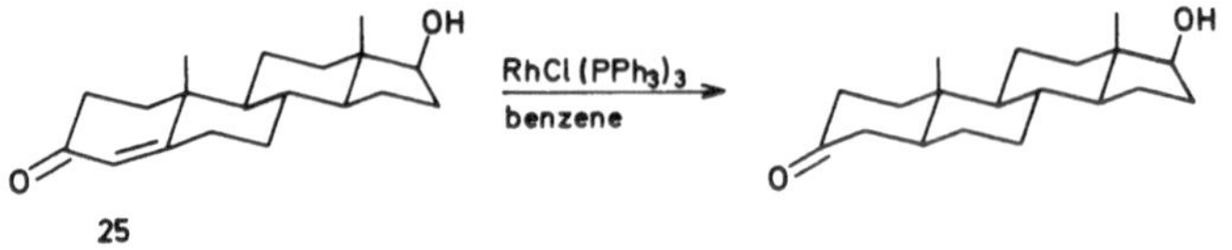

25

The selective hydrogenation of 2,3-dimethoxy-1,4-benzoquinone **(26)** and dehydrogriseofulvine[11] **(27)** should also be mentioned. Some of the easily

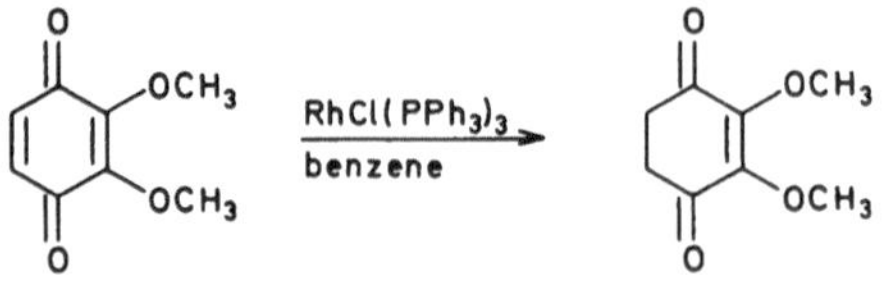

26

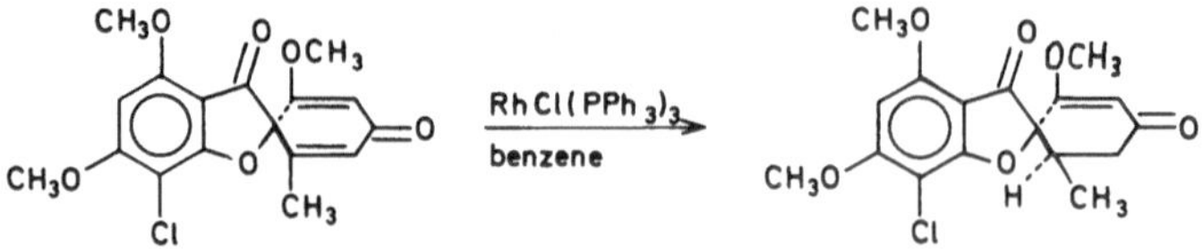

27

accessible tetrahydropyran-2-yloxycyclohexa-1,4-dienes can be similarly hydrogenated to the interesting cyclohexenyl tetrahydropyranyl ethers[11].

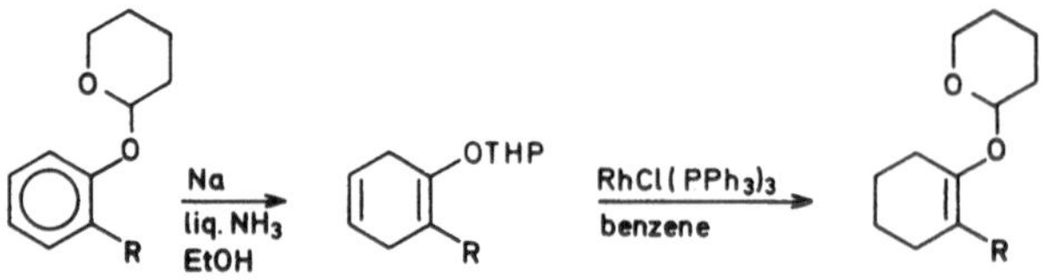

Heterogeneous hydrogenation of vinylcyclopropanes is usually accompanied by ring opening[65], which also occurs with chlorotris(triphenylphosphine)-rhodium(I) as the catalyst[72].

48

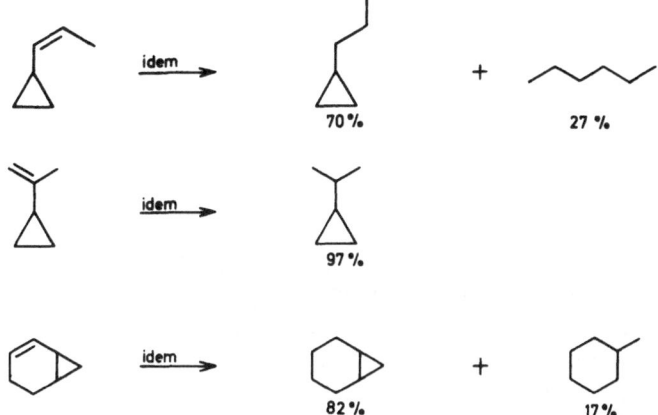

Heterogeneous hydrogenation of alkenes containing sulphur results in most cases in hydrogenolysis of the carbon-sulphur bond and concomitant poisoning of the catalyst. Chlorotris(triphenylphosphine)rhodium(I), however, is rather insensitive to poisoning by sulphur compounds such as thiophenol and alkylsulphides[73], and has been applied successfully to the hydrogenation of alkenylthiophenes and related compounds[74].

References

1. B.R. James, *Homogeneous hydrogenation,* Wiley-Interscience, London, 1973.
2. A.P.G. Kieboom and H. van Bekkum, *J. Catal.,* 25, 342 (1972).
3. D.E. Budd, D.G. Holah, A.N. Hughes, and B.C. Hui, *Can. J. Chem.,* 52, 775 (1974).
4. *Cf.* F.R. Hartley, *Chem. Rev.,* 69, 799 (1969).
5. C.A. Tolman, *J. Am. Chem. Soc.,* 96, 2780 (1974); W. Partenheimer and B. Durham, *Ibid.,* 96, 3800 (1974).
6. C.A. Brown and V.K. Ahuja, *J. Org. Chem.,* 38, 2226 (1973).
7. F.H. Jardine, J.A. Osborn, and G. Wilkinson, *J. Chem. Soc. A,* 1967, 1574.
8. P.S. Hallman, B.R. McGarvey, and G. Wilkinson, *Ibid.,* 1968, 3143.
9. C. O'Connor and G. Wilkinson, *Ibid.,* 1968, 2665.
10. A. Oudeman, F. van Rantwijk, and H. van Bekkum, *J. Coord. Chem.,* 4, 1 (1974).
11. A.J. Birch and K.A.M. Walker, *Aust. J. Chem.,* 24, 513 (1971).
12. S. Siegel, *Adv. Catal.,* 16, 123 (1966).
13. H. van Bekkum, F. van Rantwijk, G. van Minnen-Pathuis, J.D. Remijnse, and A. van Veen, *Rec. Trav. Chim. Pays-Bas,* 88, 911 (1969).

49

14. M. Peque and R. Maurel, *J. Catal.*, 19, 360 (1970).

15. a. F. van Rantwijk, A. van Vliet, and H. van Bekkum, *J. Chem. Soc., Chem. Commun.*, 1973, 234; b. F. van Rantwijk, A.P.G. Kieboom, and H. van Bekkum, in: *Catalysis*, B. Delman and G. Jannes, Eds., Elsevier Amsterdam, 1975, p 53; *J. Mol. Catal.*, 1, 27 (1975/6).

16. F. van Rantwijk, A. van Vliet, and H. van Bekkum, paper in preparation.

17. *Cf.* P.N. Rylander, *Catalytic hydrogenation over platinum metals*, Academic Press, London, 1967, p 101.

18. W.F. Johns, *J. Org. Chem.*, 31, 3780 (1966).

19. R.B. Woodward, F. Sondheimer, D. Taub, K. Heusler, and W.M. McLamore, *J. Am. Chem. Soc.*, 74, 4223 (1952).

20. For a review see: R.T. Blickenstaff, A.C. Gosh, and G.C. Wolf, *Total synthesis of steroids*, Academic Press, New York, 1974, p 20.

21. L. Velluz, G. Nominé, G. Amiard, V. Torelli, and J. Cérède, *C.R. Acad. Sci., Ser. C*, 257, 3086 (1963).

22. B. van de Graaf, H. van Bekkum, and B.M. Wepster, *Rec. Trav. Chim. Pays-Bas*, 87, 777 (1968).

23. S. Siegel and B. Dmuchovsky, *J. Am. Chem. Soc.*, 84, 3132 (1962).

24. J.F. Sauvage, R.H. Baker, and A.S. Hussey, *Ibid.*, 82, 6090 (1960).

25. H. van Bekkum, B. van de Graaf, G. van Minnen-Pathuis, J.A. Peters, and B.M. Wepster, *Rec. Trav. Chim. Pays-Bas*, 89, 521 (1970).

26. W.Th.M. de Groot, A. Steenhoek, A.M. van Wijk, and H. van Bekkum, unpublished results.

27. P.T. Alderliesten and A.P.G. Kieboom, unpublished results.

28. D. Gagnaire and P. Vottero, *Bull. Soc. Chim. Fr.*, 1970, 164.

29. *Cf.* ref. 1, pp 220, 226.

30. H.W. Thompson, *Ann. N.Y. Acad. Sci.*, 214, 195 (1973).

31. H.W. Thompson and E. McPherson, *J. Am. Chem. Soc.*, 96, 6232 (1974).

32. See pertinent references cited in ref. 15[a].

33. T.C. McKenzie, *J. Org. Chem.*, 39, 629 (1974).

34. T.R.B. Mitchell, *J. Chem. Soc. B*, 1970, 823.

35. H. van Bekkum, J. van Gogh, and G. van Minnen-Pathuis, *J. Catal.*, 7, 292 (1967).

36. F. van Rantwijk and H. van Bekkum, *J. Mol. Catal.*, 1, 383 (1975/76).

37. I. Horiuti and M. Polanyi, *Trans. Faraday Soc.*, 30, 1164 (1934); G.V. Smith and R.L. Burwell, Jr., *J. Am. Chem. Soc.*, 84, 925 (1962); R.L. Burwell, Jr. and J.B. Peri, *Annu. Rev. Phys. Chem.*, 15, 131 (1964); for alternative mechanisms see R. Touroude and F.G. Gault, *J. Catal.*, 37, 193 (1975).

38. S. Mitsui, K. Gohke, H. Saito, A. Nanbu, and Y. Senda, *Tetrahedron*, 29, 1523 (1973).

39. G.C. Bond and P.B. Wells, *Adv. Catal.*, 15, 91 (1964); G.C. Bond and J.S. Rank, *Proc. Int. Congr. Catal., 3rd., 1964*, 2, 1225 (1965).

40. G.V. Smith, J.A. Roth, D.S. Desai, and J.L. Kosco, *J. Catal.*, **30**, 79 (1973).
41. J.J. Rooney, F.G. Gault, and C. Kemball, *Ibid.*, **1**, 255 (1962); J.J. Rooney, *Ibid.*, **2**, 53 (1963).
42. R.L. Burwell, *Acc. Chem. Res.*, **2**, 289 (1969).
43. G.V. Smith and D.S. Desai, *Ann. N.Y. Acad. Sci.*, **214**, 20 (1973).
44. J.G. Atkinson and M.O. Luke, *Can. J. Chem.*, **48**, 3580 (1970).
45. E.J. Corey and J.W. Suggs, *J. Org. Chem.*, **38**, 3224 (1973); see also P. Golborn and F. Scheinmann, *J. Chem. Soc., Perkin Trans. 1*, **1973**, 2870.
46. P.A. Gent and R. Gigg, *J. Chem. Soc., Chem. Commun.*, **1974**, 277.
47. S. Siegel and G.V. Smith, *J. Am. Chem. Soc.*, **82**, 6082, 6087 (1960); S. Mitsui, S. Imaizumi, A. Nanbu, and Y. Senda, *J. Catal.*, **36**, 333 (1975).
48. S. Mitsui, Y. Senda, H. Suzuki, S. Sekiguchi, and Y. Kumagai, *Tetrahedron*, **29**, 3341 (1973).
49. S. Siegel and D.W. Ohrt, in *Catalysis*, B. Delman and G. Jannes, Eds., Elsevier, Amsterdam, 1975, p 219.
50. F. van Rantwijk, G.J. Timmermans, and H. van Bekkum, *Rec. Trav. Chim. Pays-Bas*, **95**, 39 (1976).
51. R.E. Ireland and P. Bey, *Org. Synth.*, **53**, 63 (1973).
52. *Cf.* ref. 1, p 212.
53. C.A. Brown, *J. Org. Chem.*, **35**, 1900 (1970).
54. P. Lombardi, *Gazz. Chim. Ital.*, **104**, 867 (1974).
55. K.M. Nicholas, *J. Am. Chem. Soc.*, **97**, 3254 (1975).
56. D.R. Fahey, *J. Org. Chem.*, **38**, 80 (1973).
57. C. Okkerse, A. de Jonge, J.W.E. Coenen, and A. Rozendaal, *J. Am. Oil Chem. Soc.*, **44**, 152 (1967).
58. A.P. Phillips and J. Mentha, *J. Am. Chem. Soc.*, **78**, 140 (1956).
59. T.W. Russell and D.M. Duncan, *J. Org. Chem.*, **39**, 3050 (1974).
60. A.J. Birch and K.A.M. Walker, *Tetrahedron Lett.*, **1967**, 3457.
61. F.H. Jardine and G. Wilkinson, *J. Chem. Soc. C*, **1967**, 270.
62. R. Noyori, I. Umeda, and T. Ishigami, *J. Org. Chem.*, **37**, 1542 (1972).
63. T. Kitamura, N. Sakamoto, and T. Joh, *Chem. Lett.*, **1973**, 379.
64. W.A. Bonner, N.I. Burke, W.E. Fleck, R.K. Hill, J.A. Joule, B. Sjöberg, and J.H. Zalkow, *Tetrahedron*, **20**, 1419 (1964).
65. W.T. Russell and R.H. Roy, *J. Org. Chem.*, **36**, 2018 (1971).
66. G.E. Ham and W.P. Coker, *Ibid.*, **29**, 194 (1964).
67. H.-M. Fischler, W. Hartmann, H.-G. Heine, and O. Weissel, Ger. Offen 2,163,394, *Chem. Abstr.*, **79**, 65887 (1973).
68. A.P.G. Kieboom, unpublished results.
69. R.E. Harmon, J.L. Parsons, D.W. Cooke, S.K. Gupta, and J. Schoolenberg, *J. Org. Chem.*, **34**, 3684 (1969).
70. A.J. Birch and K.A.M. Walker, *J. Chem. Soc. C*, **1966**, 1894.
71. For further examples see: F.J. McQuillin, *Prog. Org. Chem.*, **8**, 314 (1973); *Homogeneous catalysis in organic and inorganic chemistry, Vol 1:*

Homogeneous hydrogenation in organic chemistry, Reidel, Boston, 1975. A review of the hydrogenation of alkenic steroids in the presence of Wilkinson catalyst is given in ref. 1, p 226.

72. C.H. Heathcock and S.R. Poulter, *Tetrahedron Lett.*, **1969**, 2755.
73. A.J. Birch and K.A.M. Walker, *Ibid.*, **1967**, 1935.
74. A.-B. Hornfeldt, J.S. Gronowitz, and S. Gronowitz, *Acta Chem. Scand.*, **22**, 2725 (1968).
75. S. Siegel and J.R. Cozort, *J. Org. Chem.*, **40**, 3594 (1975).
76. R.R. Schrock and J.A. Osborn, *J. Am. Chem. Soc.*, **98**, 4450 (1976).

3. Hydrogenation of aromatic rings

Hydrogenation of carbocyclic arenes is usually carried out with platinum, rhodium, ruthenium, nickel or palladium as catalyst, in many cases at elevated temperatures and/or pressures. Homogeneous catalysts are seldom active for the hydrogenation of arenes. Platinum and rhodium are the preferred catalysts for the hydrogenation of pyridines and pyrroles respectively. Furans are preferably hydrogenated over palladium or rhodium since the other catalysts tend to cause ring opening. Ferrocene can be hydrogenated under carefully chosen conditions[1].

PARTIAL HYDROGENATION

The hydrogenation of an aromatic ring is best regarded as a consecutive transfer of six hydrogen atoms[2,3], various partially hydrogenated species are therefore present as surface complexes. These species are hydrogenated in competition with unconverted arene; desorption of partially hydrogenated products depends, therefore, on the position of the chemisorption equilibrium and the relative rates of hydrogen transfer.

Formation of cyclohexadienes takes place only occasionally and is generally unlikely for thermodynamic reasons. The two-step hydrogenation of [2.2]-paracyclophane constitutes an example[4].

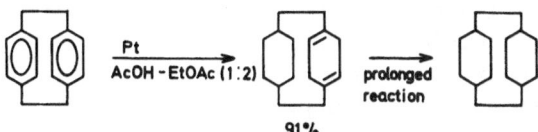

Intermediate desorption of cyclohexenes is very common, however[5]; the concentration attained in the course of the reaction depends strongly on the catalyst, the nature of the substrate and the reaction conditions, subject to the considerations mentioned above. In the case of benzene, only 1% is desorbed as cyclohexene upon hydrogenation in the liquid phase over ruthenium at 30° and 1 atm[6], whereas gas-phase hydrogenation over platinum deposited on nylon-6 has been reported to give up to 48% cyclohexene[7].

Substitution of the ring with bulky groups seems to favour intermediate desorption, especially with rhodium or ruthenium as the catalyst[8-10]. As a general rule, cyclohexene desorption is somewhat less pronounced upon hydrogenation over platinum, and still less with palladium as the catalyst[2,8-10]. Thus, upon hydrogenation of 2-t-butylbenzoic acid over rhodium, about 75% of the reactant adopted six hydrogen atoms during one residence at the catalyst surface, 22% desorbed as 6-t-butyl-1-cyclohexenecarboxylic acid which accumulated in the course of the reaction[9], and minor quantities of 2-t-butyl-1-cyclohexene-carboxylic acid were also detected. Similarly, hydrogenation of 1,2-di-t-butylbenzene over rhodium resulted in formation of up to 45% 2,3-di-t-butyl-cyclohexene[11]; 1,3,5-tri-t-butylbenzene exhibited 75% desorption of the

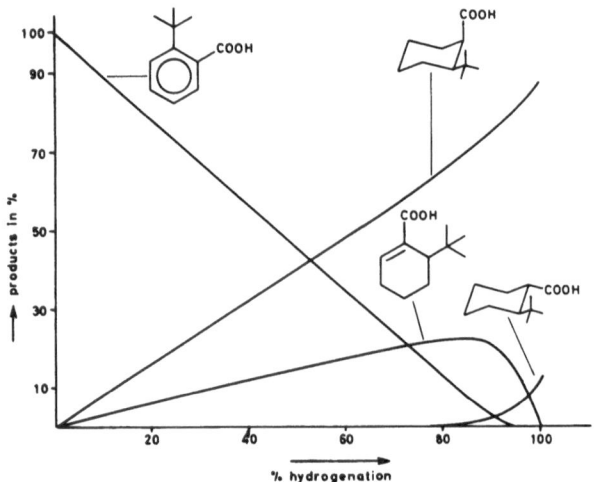

Hydrogenation of 2-t-butylbenzoic acid over Rh/C in ethanol at 22° and 1 atm.

cyclohexene in the course of hydrogenation over rhodium[10]. For a selection of reactants and catalysts a comparison is given in Table IV.

Partial hydrogenation of substituted arenes has been mainly investigated in the course of mechanistic studies but application for preparative purposes has remained rare. Partial hydrogenation of phenols occupies a somewhat special position in view of the keto-enol equilibria of partially hydrogenated products. Partial hydrogenation of phenol itself is an important step in the manufacture

54

of ε-caprolactam; typically palladium or nickel are used in aqueous alcohol at 185° and 5 atm[12]. Likewise, partial hydrogenation of resorcinol afforded cyclohexane-1,3-dione[13].

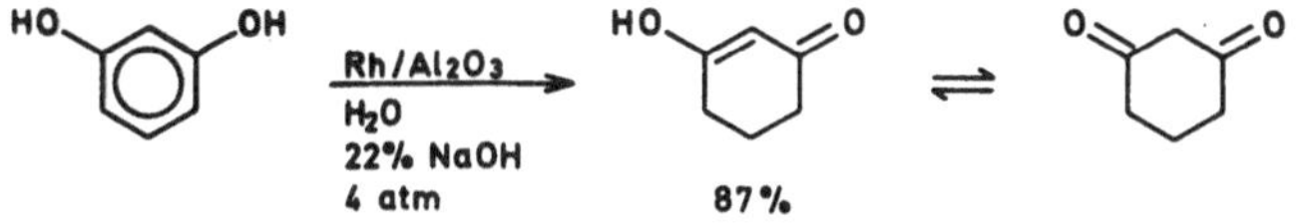

Table IV: Cyclohexene intermediates in the hydrogenation of *t*-butylbenzenes[a]

Reactant	Catalyst	Solvent	Max. % alkene	Ref.
methyl 2-*t*-butylbenzoate	Rh/C	EtOH	20	9
	Ru/C	EtOH	9	9
	Pt/C	AcOH	<0.1	9
	Pd/C	n-C$_7$H$_{16}$	<0.1	9
2-*t*-butylbenzoic acid	Rh/C	EtOH	23	9
	Rh/C	H$_2$O, OH$^-$	6	9
	Pt/C	AcOH	<0.1	9
1,2-di-*t*-butylbenzene	Rh/C	AcOH	45	11
	Rh/C	EtOH	30	11
	Pt/C	EtOH	6	11
1,3,5-tri-*t*-butylbenzene	Rh/C	n-C$_7$H$_{16}$	65	10
	Pt/C	n-C$_7$H$_{16}$	50	10
	Pd/C	n-C$_7$H$_{16}$	12	10

a: All hydrogenations were conducted under ambient conditions.

STEREOCHEMISTRY

If hydrogenation takes place during a single period of residence on the catalyst surface, suprafacial addition of six hydrogen atoms to the chemisorbed face of the ring would be expected. The stereochemical picture is considerably more varied since desorbed cyclohexene and/or roll-over mechanisms may serve as a source of overall antarafacially hydrogenated product. A striking example is found in the hydrogenation of 1,2-di-*t*-butylbenzene over rhodium[11]. It should be mentioned that the high degree of antarafacial hydrogenation of 2-*t*-butylbenzoic acid over palladium (63%), does not correspond to a high concentration of the cyclohexene (<0.1%)[9]. This may indicate rapid renewed adsorption or some alternative hydrogenation mechanism.

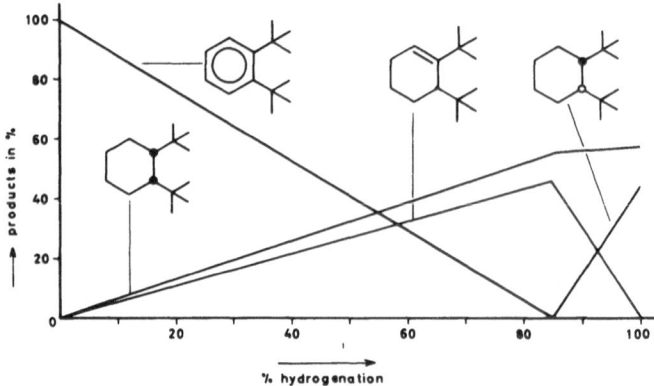

Hydrogenation of 1,2-di-t-butylbenzene over Rh/C in acetic acid at 25° and 1 atm[11].

In order to effect stereospecific suprafacial hydrogenation it is necessary to achieve transfer of six hydrogen atoms during one period of residence at the catalyst surface. Suprafacial hydrogenation of *o*-disubstituted benzenes can be effected by the use of platinum as catalyst[9,11]. Clean suprafacial hydrogenation has also been reported for the homogeneous catalyst (η^3-allyl)tris(trimethylphosphite)cobalt(I)[14]. Iridium and osmium have been recommended as efficient catalysts for the stereospecific hydrogenation of *o*-xylene (97-99% cis-1,2-dimethylcyclohexane)[15] as well as *m*- and *p*-xylene (up to 90% *cis*). Ruthenium proved to be very suitable for the stereospecific hydrogenation of alkylpyridines at 150° and 200 atm. *cis*-Dimethyl, -diethyl- and -2,4,6-trimethylpiperidine were formed in 88—100% yields[16]*.

In order to increase the probability of transfer of six hydrogen atoms during one period of residence at the catalyst surface, reaction at elevated pressure would seem advantageous. A slight increase of the *cis/trans* ratio with pressure has indeed been reported for the hydrogenation of the xylenes and 4-*t*-butyltoluene over Adams platinum[17].

Hydrogenation of some 2-alkylbiphenyls over Raney nickel at 80° and 50 atm gave the corresponding *cis*-2-alkylcyclohexylcyclohexanes in yields of 65-85%. It is interesting to note that addition of the first six hydrogen atoms proceeded with a slight preference for the substituted ring. Thus, the *cis*-2-alkylphenylcyclohexane was a major (45-63%) intermediate product[18].

---

* Surprisingly, the stereospecific hydrogenation of the thiophene diurethane **9** (the key step in a recent synthesis of biotin) was achieved with palladium as the catalyst[32].

CH₃OOCN   NCOOC₂H₅   CH₃OOCN   NCOOC₂H₅
                          Pd/C
         (CH₂)₄COOH   ———————→   (CH₂)₄COOH
                         AcOH
                         124 at, 50°
         9   95%

The course of the hydrogenation of a number of trypticene derivatives has been investigated[19]. Hydrogenation of trypticene itself (1) over palladium or ruthenium at 100-200° and 30-100 atm gave up to 60% of all-*cis-anti*-octadeca-hydrotrypticene (2), with prolonged reaction times isomerization to give the all-*trans*-isomer (3) occurred[20].

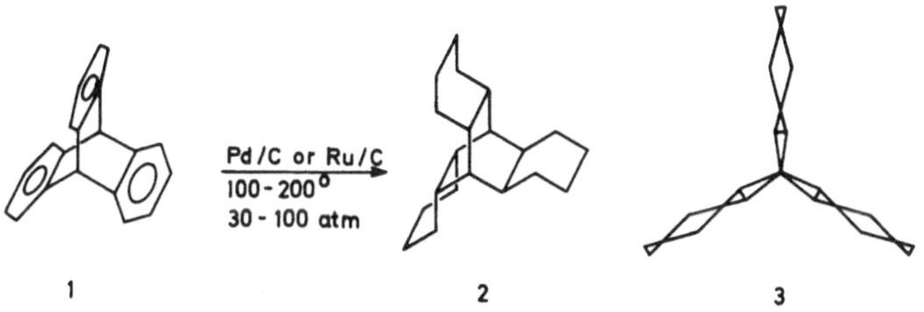

Spectacular results have been obtained with the homogeneous catalyst (η^3-allyl)tris(trimethylphosphite)cobalt(I)[14]. Complete stereospecificity with this catalyst has been demonstrated for a number of dialkylbenzenes[14a], hydrogenation of hexadeuterobenzene afforded all-*cis*-hexadeuterocyclohexane[14c].

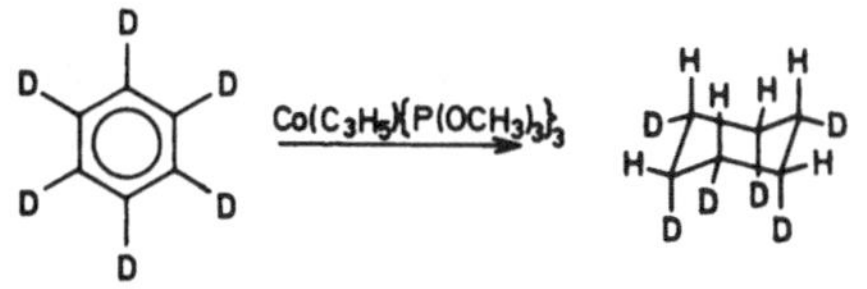

Nickelocene (4) also gave interesting results: reaction with hydrogen at 50° and 30 atm gave (η^3-cyclopentenyl)(η^5-cyclopentadienyl)nickel(II); in the presence of Raney nickel, palladium or chlorotris(triphenylphosphine)rhodium(I) milder conditions could be applied. In all cases hydrogen entered from the side of the nickel atom, suggesting participation of nickel in the hydrogen transfer step[21].

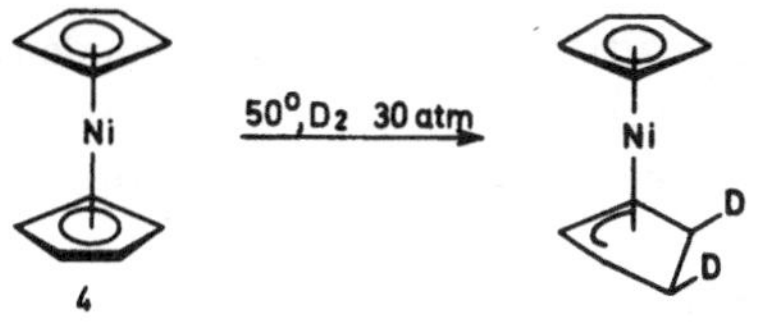

PARTIAL HYDROGENATION OF NAPHTHALENES

The hydrogenation of naphthalenes proceeds in two distinct stages. Initially a tetralin is formed, this is then further hydrogenated to the decalin; because of preferential adsorption and the faster rate of hydrogenation of the naphthalene there is effectively no hydrogenation of the tetralin until all the naphthalene has been used up[22]. In the hydrogenation of 2-alkylnaphthalenes a slight

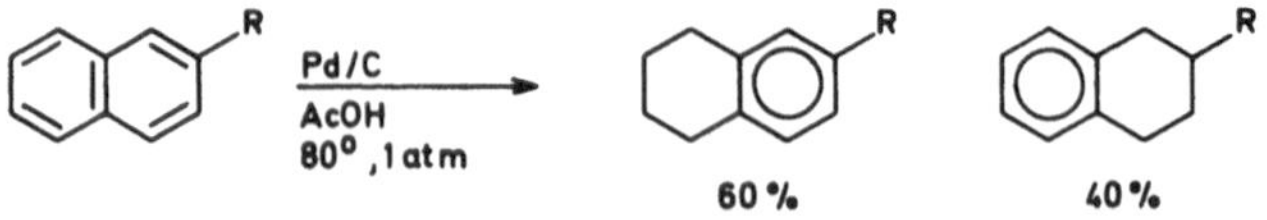

| | 60 % | 40 % |

preference for saturation of the unsubstituted ring has been established[23]. The regioselectivity of the hydrogenation of 1-alkylnaphthalenes follows an interes-ting trend: with small substituents hydrogenation of the unsubstituted ring is favoured, whereas bulky groups promote saturation of the substituted ring[23]

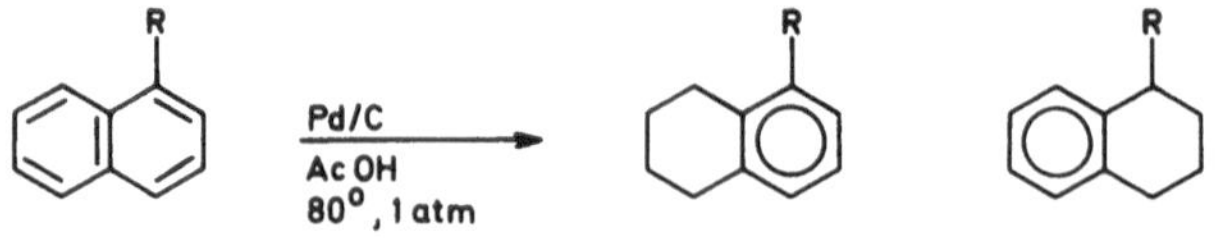

R = Me	66 %	34 %
Et	55 %	45 %
i-Pr	32 %	68 %
t-Bu	3 %	97 %

The regioselectivity can be analyzed in terms of steric approach control favouring the unsubstituted ring, and steric development control, favouring the substituted ring due to relief of peri-strain[23]. 1,8-Dialkylnaphthalenes reacted rapidly compared with related compounds, this effect is also due to relief of peri-strain[23].

From a preparative point of view the regioselectivity exhibited in the palladium catalyzed hydrogenation of 1-t-butylnaphthalene (5) contrasts nicely with the results of reduction by lithium in liquid ammonia[23,24].

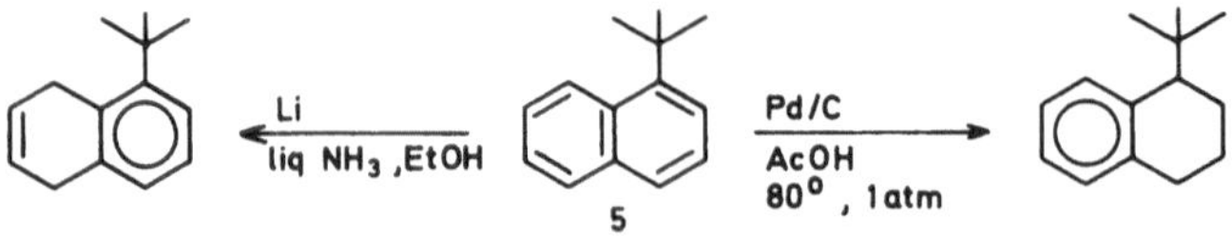

An example of homogeneously catalyzed partial hydrogenation of condensed arenes is found in the reduction of 9,10-dimethylanthracene in the presence of cobalt carbonyl complexes. Mixtures of cis- and trans-9,10-dihydro-9,10 dimethylanthracene were formed[25].

It is pertinent to note that selective partial hydrogenation of the benzene ring in quinoline, using platinum oxide in the presence of strong acid has recently been reported[31].

REDUCIBLE FUNCTIONS

Unsaturated functions are usually not preserved under the conditions of arene hydrogenation. Nitro and halogen functions are usually reduced, esters are not attacked. The solvent is sometimes important, as exemplified in the case of o-methoxyphenylpropanone (6). On hydrogenation over rhodium in acetic acid 2-methoxycyclohexylpropanone is obtained, whilst in methanol the carbonyl function is also attacked[26].

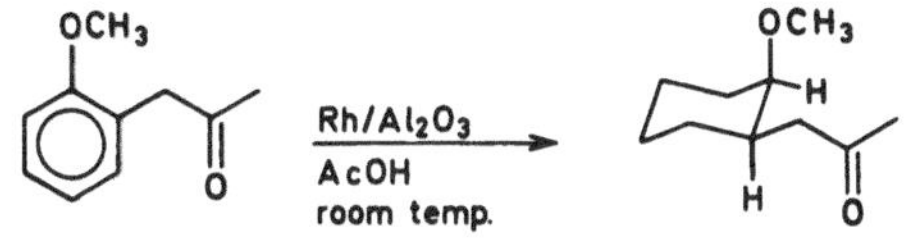

Benzyl derivatives are extremely sensitive and should be treated with special care. The tendency to effect hydrogenolysis increases in the order ruthenium < rhodium < platinum < palladium. With palladium as the catalyst, benzyl hydrogenolysis generally proceeds faster than hydrogenation of the aromatic ring. Selective hydrogenation of benzyl derivatives is often possible, however, under carefully chosen conditions. The selective hydrogenation of 4,4'-bis(aminomethyl)bibenzyl (7) over ruthenium is an example[27a]. It is worth noting — *inter alia* — that the isomer distribution was rather temperature-sensitive[27b].

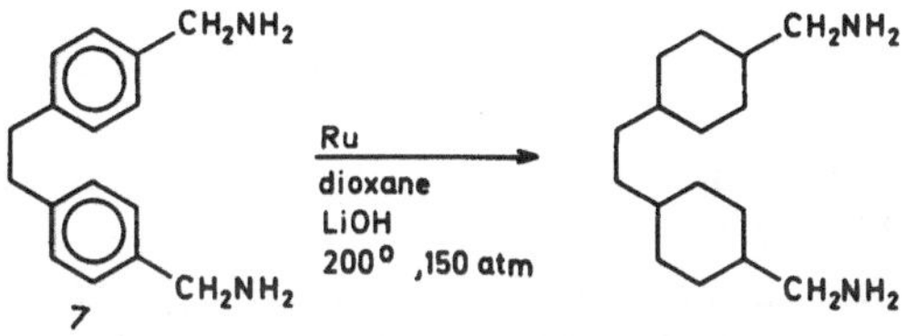

The use of rhodium or iridium is to be recommended for the selective hydrogenation of phenols and phenyl ethers. Thus, hydrogenation of 1-naphthol over rhodium at 25° and 130 atm gave 1-decalol[28], and in a similar way cyclohexane-r-1,c-2,c-3-triol was obtained from pyrogallol[29]. Selective hydrogenation of dibenzo-18-crown-6 polyether (8) could only be achieved

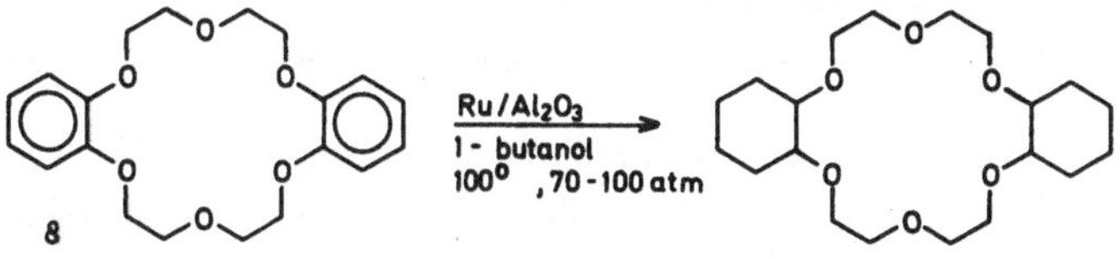

with ruthenium as the catalyst[30]. It should be mentioned that rhodium and ruthenium are also the preferred catalysts for the hydrogenation of anilines.

SIDE-REACTIONS

The partial hydrogenation of phenols to give cyclohexanones has already been mentioned. In a similar fashion, hydrogenation of anilines proceeds *via* iminocyclohexanes; coupling and hydrogenolysis may then give rise to dicyclo- hexylamines (*cf.* p 77). Inhibition of reductive coupling is possible by a suitable choise of conditions, as well as by hydrogenation in the presence of ammonia.

References

1. F. van Meurs, F.W. Metselaar, A.J.A. Post, J.A.A.M. van Rossum, A.M. van Wijk, and H. van Bekkum, *J. Organomet. Chem.*, **84**, C22 (1975).
2. For a review see: S. Siegel, *Adv. Catal.*, **16**, 123 (1966).
3. D. Shopov and A. Andreev, *J. Catal.*, **6**, 316 (1966); J. Völter, M. Hermann, and K. Heise, *Ibid.*, **12**, 307 (1968).
4. D.J. Cram and N.L. Allinger, *J. Am. Chem. Soc.*, **77**, 6289 (1955).
5. H. van Bekkum, Thesis, Delft University of Technology, The Netherlands, 1970, pp 9–14.
6. F. Hartog, J.H. Tebben, and C.A.M. Weterings, *Proc. Int. Congr. Catal. 3rd, 1964*, **2**, 1210 (1965).
7. P. Dini, D. Dones, S. Montelatici, and N. Giordano, *J. Catal.*, **30**, 1 (1973).
8. F. Hartog and P. Zwietering, *Ibid.*, **2**, 79 (1963).
9. H. van Bekkum, B. van de Graaf, G. van Minnen-Pathuis, J.A. Peters, and B.M. Wepster, *Rec. Trav. Chim. Pays-Bas*, **89**, 521 (1970).
10. H. van Bekkum, H.M.A. Buurmans, G. van Minnen-Pathuis, and B.M. Wepster, *Ibid.*, **88**, 779 (1969).
11. B. van de Graaf, H. van Bekkum, and B.M. Wepster, *Ibid.*, **87**, 777 (1968).
12. R.J. Duggan, E.J. Murray, and L.O. Windstrom, U.S. patent 3,076,810 (1963); *Chem. Abstr.*, **59**, 2671h (1963).
13. H.A. Smith and B.L. Stump, *J. Am. Chem. Soc.*, **83**, 2739 (1961).
14. *a.* E.L. Muetterties and F.J. Hirsekorn, *Ibid.*, **96**, 4063 (1974); *b.* F.J. Hirsekorn, M.C. Rakowski, and E.L. Muetterties, *Ibid.*, **97**, 237 (1975); *c.* E.L. Muetterties, M.C. Rakowski, F.J. Hirsekorn, W.D. Larson, V.J. Basus, and F.A.L. Anet, *Ibid.*, **97**, 1266 (1975).
15. S. Nishimura, F. Mochizuki, and S. Kobayakawa, *Bull. Chem. Soc. Jpn.*, **43**, 1919 (1970).
16. R. Schubart, H. Ziemann, and D. Wendisch, *Synthesis*, **1973**, 220.
17. S. Siegel, G.V. Smith, B. Dmuchovsky, D. Dubbell, and W. Halpern, *J. Am. Chem. Soc.*, **84**, 3136 (1962).
18. G. Descotes, P. Legrand-Berlebach, and J. Sabadie, *Bull. Soc. Chim. Fr.*, **1973**, 1517.
19. V.L. Mylroie and J.F. Stenberg, *Ann. N.Y. Acad. Sci.*, **214**, 255 (1973).
20. C. Morandi, E. Mantica, D. Botta, M.T. Gramegna, and M. Farina, *Tetra- hedron Lett.*, **1973**, 1141.
21. K.W. Barnett, F.D. Mango, and C.A. Reilly, *J. Am. Chem. Soc.*, **91**, 3387 (1969).

22. Th. J. Nieuwstad, Thesis, Delft University of Technology, 1970.
23. Th. J. Nieuwstad, P. Klapwijk, and H. van Bekkum, *J. Catal.*, **29**, 404 (1973).
24. Th. J. Nieuwstad and H. van Bekkum, *Rec. Trav. Chim. Pays-Bas*, **91**, 1069 (1972).
25. P.D. Taylor and M. Orchin, *J. Org. Chem.*, **37**, 3913 (1972).
26. S.E. Cantor and D.S. Tarbell, *J. Am. Chem. Soc.*, **86**, 2902 (1964).
27. H. van Brederode, Thesis, Delft University of Technology, 1975, *a.* p 62; *b.* p 72.
28. M. Freifelder and G.R. Stone, *J. Pharm. Sci.*, **53**, 1134 (1964).
29. W.R. Christian, C.J. Gogek, and C.B. Purves, *Can. J. Chem.*, **29**, 911 (1951); *cf.* S.J. Angyal and D.J. McHugh, *J. Chem. Soc.*, **1957**, 3682.
30. C.J. Petersen, *J. Am. Chem. Soc.*, **89**, 7017 (1967).
31. F.W. Vierhapper and E.L. Eliel, *J. Org. Chem.*, **40**, 2729 (1975).
32. P.N. Confalone, G. Pizzolato, and M.R. Uskokoviè, *Helv. Chim. Acta*, **59**, 1005 (1976).

4. Semi-hydrogenation of carbon-carbon triple bonds[1]

Catalytic hydrogenation of acetylenes is usually carried out with the object of introducing a *cis* double bond (= semi-hydrogenation). Initial formation of *cis*-alkene from transfer of two hydrogen atoms from the catalyst surface to chemisorbed alkyne is evidently expected. It is obvious, however, that subsequent hydrogenation to the alkane, as well as *cis-trans* and positional isomerization are bound to occur in the absence of special precautions. From a preparative point of view, catalytic semi-hydrogenation of acetylenes is complementary to metal-ammonia reduction, which yields the *trans*-alkene.

The rates of the various surface reactions of chemisorbed alkynes and alkenes are of the same order of magnitude. Monopolization of the catalyst surface by alkyne is therefore required in order to effect selective semi-hydrogenation[2]. The chemisorption equilibrium should favour the alkyne over its reaction product and also expulsion of *cis*-alkene from the surface should proceed faster than the consecutive reactions. The tendency of the triple bond to monopolize the catalyst surface has been confirmed experimentally[3].

THE CATALYST

As to the metal, Sabatier[4] reported selective partial hydrogenation of ethyne over palladium and nickel, whereas reaction in the presence of platinum proceeded aselectively. Palladium has remained the preferred semi-hydrogenation catalyst; Raney nickel, iron and rhodium have been applied incidentally, but usually with inferior results[5]. Latterly, a semi-hydrogenation catalyst based on P-2 nickel has been reported[6]. Semi-hydrogenation of alkynes with homogeneous rhodium[30;31] and palladium[32] catalysts has been reported.

Palladium has been applied as a dispersion on carbon, calcium carbonate, barium carbonate or sulphate. The choice of carrier seems to reflect personal preferences[7], as the effect of the support has not been subjected to systematic study. Partial deactivation of the catalyst is a routine precaution against over-hydrogenation and isomerization. Since the hydrogenation of the alkyne and the isomerization or hydrogenation of the product *cis*-alkene do not occur during a single period of residence on the surface, the object of the deactivating poison is to compete with alkene for the catalyst surface, thus excluding the *cis*-alkene[8]. For that reason also a clear end-point is obtained in the presence of poison. Quinoline is commonly used for the deactivation of palladium catalyst[8-10]; potassium hydroxide has been used in the semi-hydrogenation of alkynic *tertiary* alcohols[11] in order to prevent hydrogenolysis of the C-O bond.

Palladium on calcium carbonate, alloyed with lead and poisoned with quinoline – the Lindlar catalyst[10] – has found wide application as semi-hydrogenation catalyst. Preparation of the support under carefully controlled conditions constitutes a recent improvement[8]. The resulting catalyst, which does not require alloying with lead, is probably the best universal semi-hydrogenation catalyst available.

The solvent also influences the result: the highest selectivity is achieved on hydrogenation in light petroleum or ethyl acetate, reaction in the lower alcohols leads to a decrease in selectivity[8].

MONOYNES

Semi-hydrogenation of isolated triple bonds usually proceeds without difficulty, the selectivity is in the range of 95-97%. It is interesting to note that gas-phase hydrogenation of 2-butyne over palladium on alumina at 27° gave 99% *cis*-butene[12]. Selective hydrogenation of 1-alkynes to 1-alkenes in the presence of a homogeneous carboxylatorhodium(I) catalyst has recently been reported[31].

Semi-hydrogenation has been applied successfully to the synthesis of medium-sized *cis*-cycloalkenes: *cis*-cyclononene and -decene were formed in yields of 95-97% on hydrogenation of the corresponding cycloalkyne over Lindlar catalyst[13]. Similarly, semi-hydrogenation of 1,7-cyclododecadiyne over palladium on barium sulphate afforded 87% *cis,cis*-cyclododecadiene[14].

Conjugated alkynes often give lower yields of *cis*-alkene. Thus, semi-hydrogenation of 1,3-diphenylpropyne over Lindlar catalyst proceeded with only 85% selectivity[15]; monopolization of the catalyst surface by alkyne seems not to occur in this case. Similarly, hydrogenation of methyl phenylethynecarboxylate over palladium on calcium carbonate[8] in the presence of quinoline gave only 89% methyl *cis*-2-phenylethenecarboxylate[16]; with P-2 nickel in the presence of 1,2-diaminoethane less than 75% selectivity was achieved[16]. In this particular case the homogeneous platinum/tin chloride complex catalyst[17] proved superior with a selectivity of 97%[16]. It should be noted, however, that semi-hydrogenation of various substituted methyl phenylethynecarboxylates over palladium on calcium carbonate proceeded with good results (>97% *cis*-alkene[16]).

Ethyl 1-propynecarboxylate and 1-phenylpropyne are further examples of conjugated alkynes. In both cases the Lindlar catalyst as well as the platinum/ tin complex catalyst performed badly (<80% *cis*-alkene[16]); 1-phenylpropyne was, however, semi-hydrogenated with a selectivity of 96% over P-2 nickel in the presence of 1,2-diaminoethane[6].

POLYYNES AND ENYNES

Semi-hydrogenation of a polyyne proceeds to the *cis*-polyene without selective formation of the enynes, provided that the reactivity of the different triple bonds is of the same order of magnitude. A considerable amount of polyyne is converted to the *cis*-polyene during a single period of residence on the surface[8].

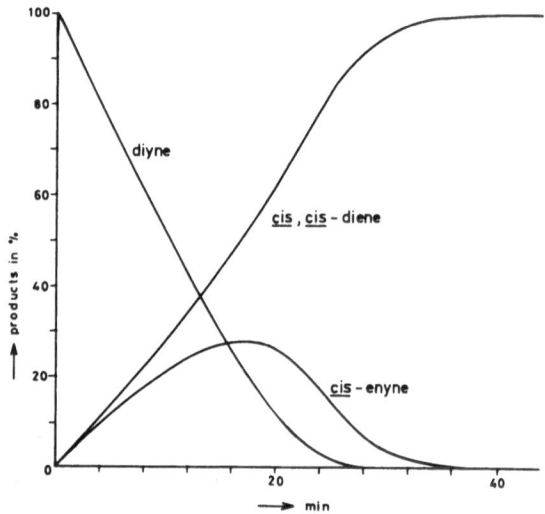

Hydrogenation of octadeca-9,12-diynecarboxylic acid over Pd/CaCO₃ in light petroleum in the presence of quinoline at 25° and 1 atm.

The semi-hydrogenation of octadeca-9,12-diynecarboxylic acid is an example: the concentration of enynes did not exceed 30% and diene formation was immediate; with nonadeca-7,10,12-triynecarboxylic acid a similar result was obtained[8].

The selectivity is generally lower with conjugated polyynes or enynes. Thus, semi-hydrogenation of 1-ethynylcyclohexene resulted in 86% formation of vinyl-cyclohexene; a terminal double bond in the presence of an internal triple bond gives rise to even greater difficulties[18]. Contrarily, the preparation of 15,15'-*cis*-β-carotene (1) according to Isler's adaptation[19] of Inhoffen's β-carotene synthesis[20] proceeded with 95% selectivity (*cf.* Scheme). The carotene synthesis also constitutes a striking example of the use of the acetylene unit as a 'linch-pin'[21]. A similar strategy was developed in the course of the annulene synthesis by Sondheimer: 1,5-hexadiyne was used as a building block for the

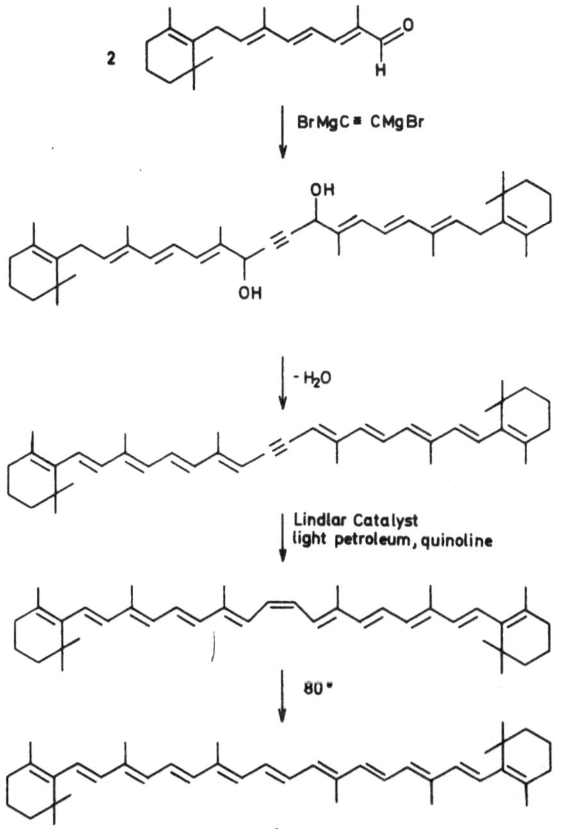

18-membered ring, which afforded [18]-annulene (2) upon isomerization followed by semi-hydrogenation[22].

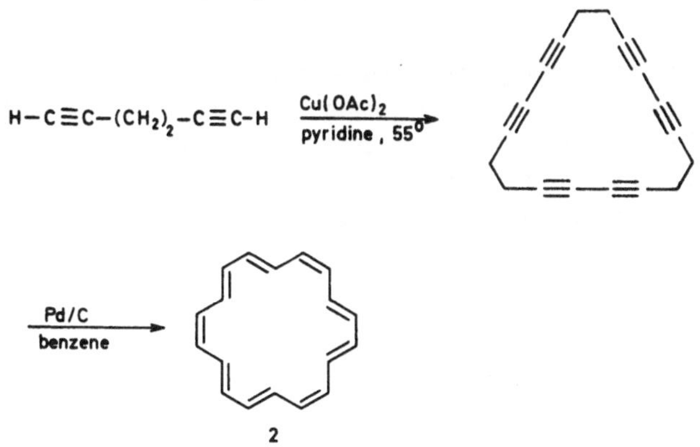

Functional groups usually remain intact. Semi-hydrogenation of *trans*-3 methoxycarbonylmethyl-2-(2-pentynyl)cyclopentanone over Lindlar catalyst afforded methyl jasmonate (3) with 94% selectivity[23]. Selective hydrogenation

3

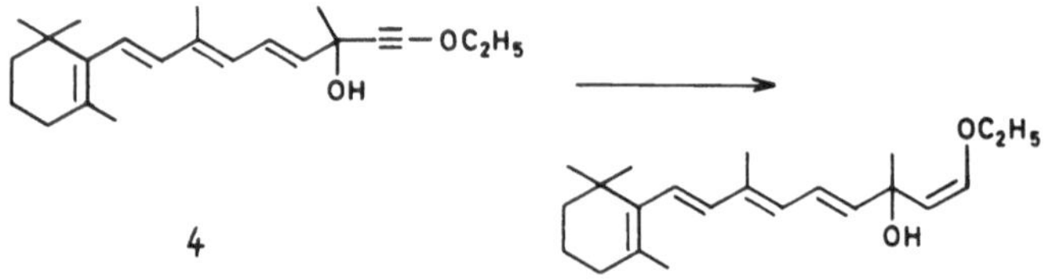

4

of ethynyl ethers such as 4 and propargyl alcohols and ethers proceeds without special problems. The reaction sequence depicted below has received wide application as a procedure for the preparation of unsaturated aldehydes[24], such as vitamin A aldehyde[25]. Various carotene and lycopene precursors (*e.g.* 5, 6

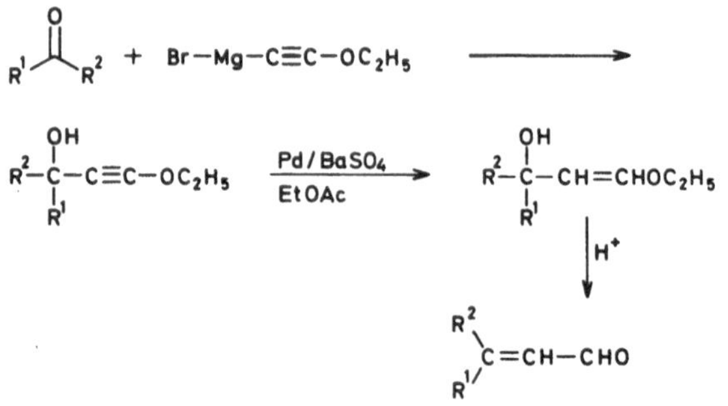

7, and 8) have also been semi-hydrogenated over palladium on carbon with quinoline[9,26], palladium on calcium carbonate[27] and palladium on barium sulphate[28]; the yields are unspecified but probably low. Hydrogenolysis of *tertiary* alcohols requires protonation of the hydroxyl group; the reaction occurs fairly easily over palladium but can be suppressed by the addition of potassium hydroxide[11,29].

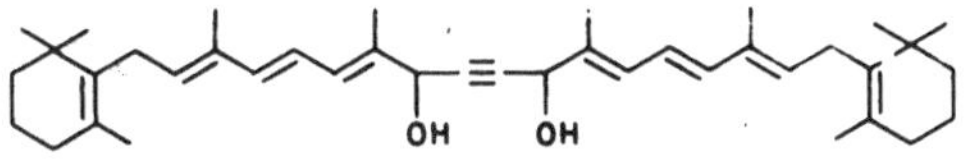

5

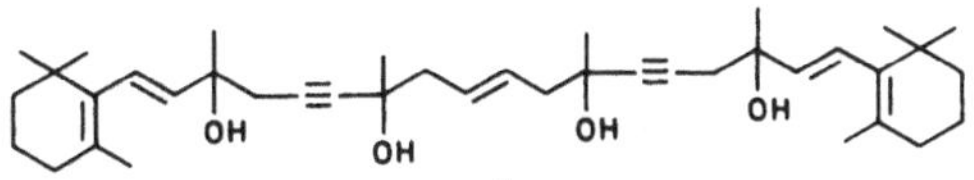

6

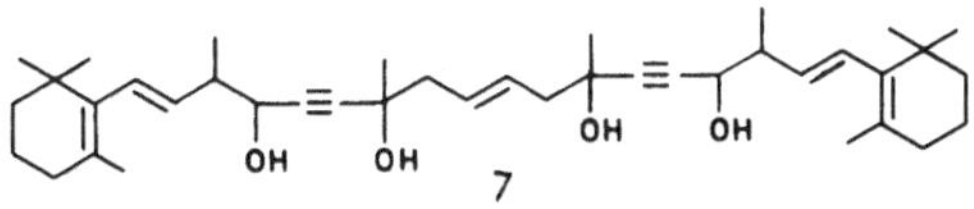

7

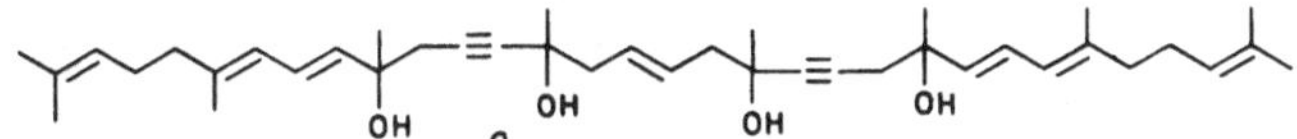

8

References

1. For a review see: E.N. Marvell and T. Li, *Synthesis*, 1973, 457.
2. A. Farkas, *Trans. Faraday Soc.*, 35, 906 (1939).
3. *Cf.* I.S. Zalkind, *Zh. Fiz. Khim.*, 52, 191 (1920), *Chem. Abstr.*, 17, 1453 (1923); M. Bourguel and V. Gredy, *C. R. Acad. Sci., Ser. C*, 189, 757 (1929); M. Bourguel, *Bull. Soc. Chim.* Fr., [4] 51, 253 (1923); Y.S. Zalkind, M.N. Vishnyakov, and L.N. Morev, *Zh. Obsch. Khim.*, 3, 91 (1933), *Chem. Abstr.* 28, 1669[5] (1934); N.A. Dobson, G. Eglinton, M. Krishnamurti, R.A. Raphael, and R.G. Willis, *Tetrahedron*, 16, 16 (1961).
4. P. Sabatier and J.B. Senderens, *C. R. Acad. Sci., Ser. C*, 128, 1173 (1899) and 131, 40 (1900).
5. Ref. 1, chapter 3 and 4, and references cited therein.
6. C.A. Brown and V.K. Ahuja, *J. Chem. Soc., Chem. Commun.*, 1973, 553.
7. In accordance with the adage: *Catalysis is an art, catalyst making is a black art.*
8. A. Steenhoek, B.H. van Wijngaarden, and H.J.J. Pabon, *Rec. Trav. Chim. Pays-Bas*, 90, 961 (1971).
9. O. Ischler, W. Huber, A. Ronco, and M. Kofler, *Helv. Chim. Acta*, 30, 1911 (1947).
10. H. Lindlar and R. Dubuis, *Org. Synth.*, Coll. Vol. 5, 880 (1973).
11. R.J. Tedeschi and G. Clark, Jr., *J. Org. Chem.*, 27, 4323 (1962).
12. W.M. Hamilton and R.L. Burwell, *Proc. Int. Congr. Catal. 2nd, 1960*, 1, 987 (1961).
13. V. Prelog, K. Schenker, and H.H. Günthard, *Helv. Chim. Acta*, 35, 1598 (1952); V. Prelog, K. Schenker, and W. Küng, *Ibid.*, 36, 471 (1953).
14. D.J. Cram and N.L. Allinger, *J. Am. Chem. Soc.*, 78, 2518 (1956).
15. E.K. Raunio and W.A. Bonner, *J. Org. Chem.*, 31, 396 (1966).

67

16. F. van Rantwijk and H. van Bekkum, paper in preparation.
17. *Cf.* H. van Bekkum, F. van Rantwijk, G. van Minnen-Pathuis, J.D. Remijnse, and A. van Veen, *Rec. Trav. Chim. Pays-Bas,* **88**, 911 (1969), and pertinent references cited therein.
18. E.N. Marvell and J. Tashiro, *J. Org. Chem.,* **30**, 3991 (1965).
19. O. Isler, H. Lindlar, M. Montavon, R. Rüegg, and P. Zeller, *Helv. Chim. Acta,* **39**, 249 (1956).
20. H.H. Inhoffen, F. Bohlmann, K. Bartram, G. Rummert, and H. Pommer, *Ann. Chem.,* **570**, 54 (1950).
21. R.A. Raphael, *Acetylene in organic synthesis,* Academic Press, New York, 1955, p 56.
22. F. Sondheimer, R. Wolovsky, and Y. Amiel, *J. Am. Chem. Soc.,* **84**, 274 (1962).
23. G. Büchi and B. Egger, *J. Org. Chem.,* **36**, 2021 (1971).
24. J.F. Arens and D.A. van Dorp, *Rec. Trav. Chim. Pays-Bas,* **67**, 973 (1948).
25. J.F. Arens and D.A. van Dorp, *Ibid.,* **68**, 604 (1949); W. Graham, D.A. van Dorp, and J.F. Arens, *Ibid.,* **68**, 609 (1949).
26. H.H. Inhoffen, F. Bohlmann, and G. Rummert, *Justus Liebigs Ann. Chem.,* **569**, 226 (1950).
27. N.A. Milas, P. Davis, I. Belič, and D.A. Fleš, *J. Am. Chem. Soc.,* **72**, 4844 (1950).
28. P. Karrer and C.H. Eugster, *Helv. Chim. Acta,* **33**, 1172 (1950); P. Karrer, C.H. Eugster, and E. Tobler, *Ibid.,* **33**, 1349 (1950).
29. R.J. Tedeschi, *J. Org. Chem.,* **27**, 2398 (1962).
30. R.R. Schrock and J.A. Osborn, *J. Am. Chem. Soc.,* **98**, 2143 (1976).
31. R.H. Crabtree, *J. Chem. Soc., Chem. Commun.,* **1975**, 647.
32. A. Sisak and F. Ungvàry, *Chem. Ber.,* **109**, 531 (1976).

5. Hydrogenation of carbon-oxygen double bonds

Catalytic reduction of aldehydes and ketones to the corresponding alcohols proceeds readily in the presence of various heterogeneous and homogeneous catalysts, usually under ambient conditions. Aliphatic aldehydes and ketones on one hand, and aromatic ones on the other, demand a somewhat different treatment and will therefore be discussed separately.

ALIPHATIC ALDEHYDES AND KETONES

Aliphatic aldehydes are hydrogenated over platinum, rhodium or ruthenium, whereas ketones react somewhat more sluggishly. Raney nickel has also found application, but palladium is much less active as a catalyst. The corresponding alcohol is mainly formed; subsequent hydrogenolysis does not constitute a serious problem.

Hydrogenation of aldehydes over platinum proceeds with concomitant deactivation of the catalyst; catalyst activity can be restored by purging with oxygen[1]. In the presence of tin(II) or iron(III) chloride, deactivation remains absent[2]. It seems likely that deactivation is caused by decarbonylation of the aldehyde and poisoning of the catalyst by carbon monoxide[3]; decarbonylation of aldehydes by metal complexes is a well-known reaction. Oxidation of chemisorbed carbon monoxide would then restore the catalyst activity. The role of the metal salts remains obscure, but partial poisoning of the catalyst surface seems plausible, attachment to the carbonyl function also seems possible. With regard to tin(II) chloride, a comparison with the platinum(II)-tin(II) homogeneous hydrogenation catalyst[4] is obvious. The hydrogenation of ketones is not subject to catalyst deactivation.

Small scale hydrogenation of simple aldehydes and ketones has now largely been superseded by reduction with hydride agents. Hydrogenation of glucose (1) over nickel to sorbitol is an example of industrial aldehyde hydrogenation[5]. A process comprising hydrogenation of wood polysaccharides under hydrolytic conditions has been described[6]. Examples of simple aldehyde and ketone hydrogenation for preparative purposes are the hydrogenation of 3-diethyl-amino-2,2-dimethylpropanal[7] (2) and 5-hydroxy-3-methylpentanal[8] over Raney nickel, and of 1,1,3,3-tetramethylcyclobutanedione (3) over ruthe-nium[9].

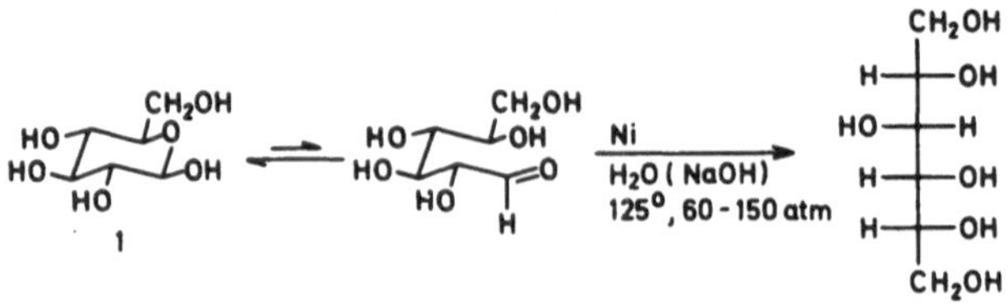

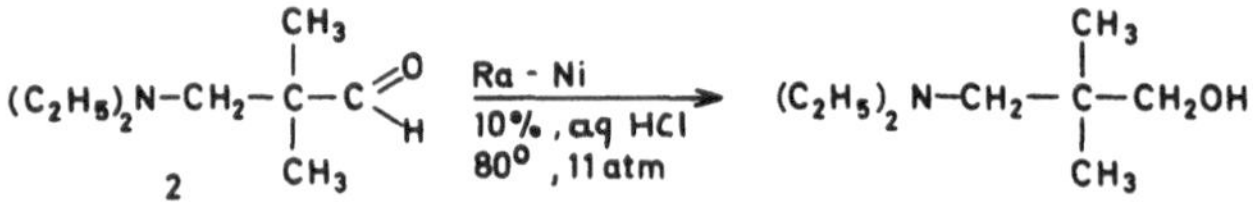

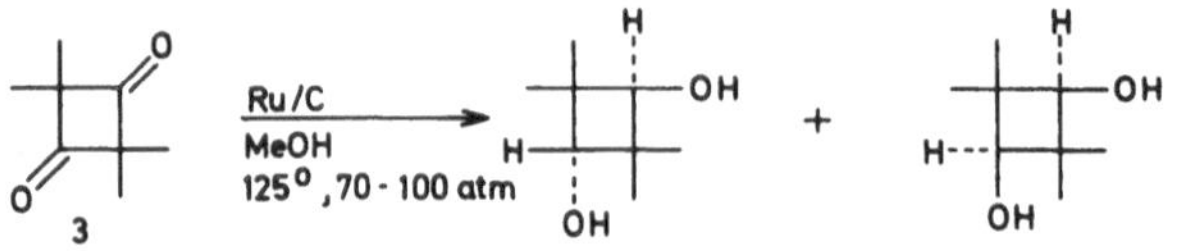

AROMATIC ALDEHYDES AND KETONES

Aromatic aldehydes and ketones are best hydrogenated over palladium, Raney nickel is somewhat less active. With palladium as the catalyst, hydrogenolysis of the benzyl alcohol formed should be taken into account. Mild hydrogenation conditions and interruption of the reaction after the consumption of one equivalent of hydrogen minimize the danger of hydrogenolysis. The effect of the temperature is shown in the hydrogenation of 3-methyl-2-carboxymethyl-1 tetralone[10] (4). With sensitive products, hydrogenolysis can be suppressed by

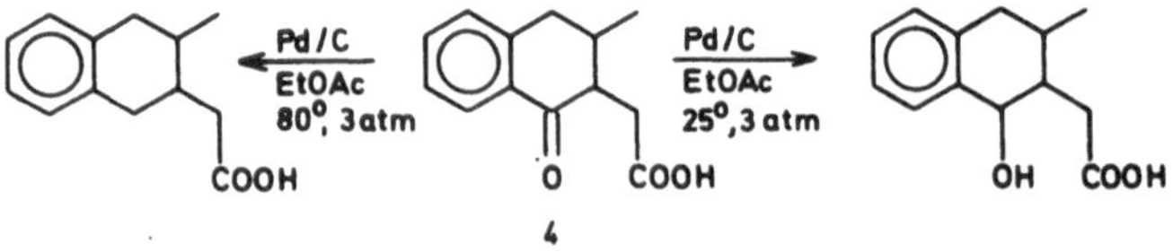

the addition of small amounts of base; accordingly, the hydrogenation of propiophenone proceeded selectively in the presence of nicotinic acid diethyl lamide[11]. Hydrogenation of 1,3,5-triacetylbenzene to 1,3,5-tris(1-hydroxy ethyl)benzene over palladium similarly proceeded selectively in the presence of amine[13].

70

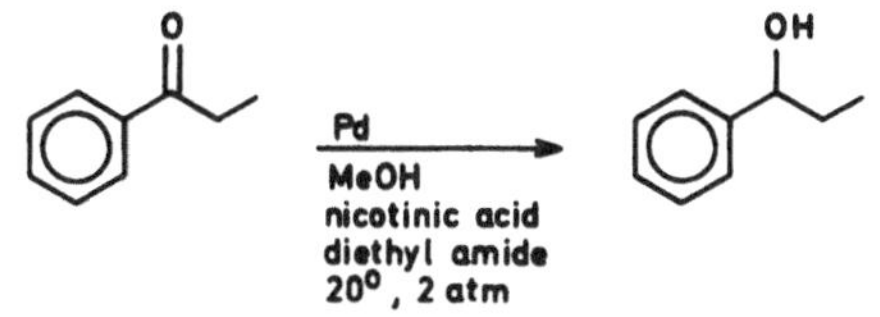

STEREOCHEMISTRY

The steric course of ketone hydrogenation is much more complicated than is the case with alkenes. According to the rules of Von Auwers[14] and Skita[15], hydrogenation of cyclohexanones over platinum would under acidic conditions result in predominant formation of the *cis*-cyclohexanol, whereas in neutral or basic solutions the *trans*-isomer would mainly be formed. Barton[16] summarized the results obtained with steroidal ketones as follows: formation of the axial alcohol predominates in acidic medium, the equatorial alcohol in neutral or basic solution. A recent electronic interpretation[17] reached a similar conclusion.

Recent work by Mitsui *et al.*[18], however, clearly showed that the steric course of hydrogenation with substituted cyclohexanones is much less straightforward. Especially striking results, which are summarized below, were obtained upon hydrogenation of 3- and 4-*t*-butylcyclohexanone (5 and 6).

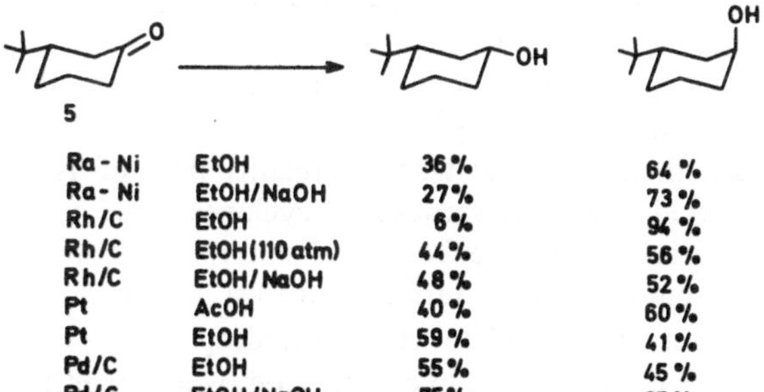

Ra - Ni	EtOH	36%	64 %
Ra- Ni	EtOH/NaOH	27%	73%
Rh/C	EtOH	6%	94 %
Rh/C	EtOH(110atm)	44%	56 %
Rh/C	EtOH/NaOH	48%	52%
Pt	AcOH	40%	60%
Pt	EtOH	59%	41%
Pd/C	EtOH	55%	45%
Pd/C	EtOH/NaOH	75%	25%

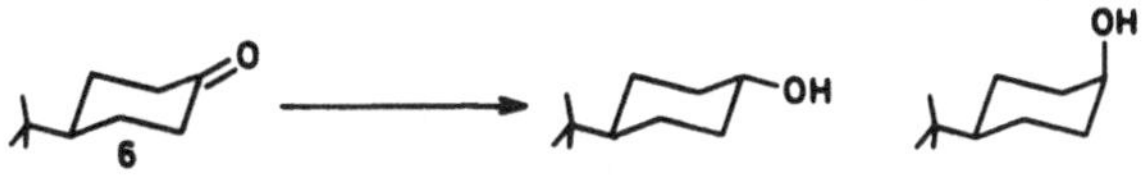

Ra - Ni	EtOH	26 %	74%
Ra- Ni	EtOH/NaOH	12%	88%
Rh/C	EtOH	6%	94 %
Rh/C	EtOH (80atm)	42%	58%
Rh/C	EtOH/NaOH	37%	63%
Pt	AcOH	32%	68%
Pt	EtOH	43%	57%
Pd/C	EtOH	56%	44%
Pd/C	EtOH/NaOH	77%	23%

The rules of Auwers-Skita and Barton were inverted for Raney nickel; palladium under alkaline conditions gave the highest yields of equatorial alcohol, whilst rhodium in neutral medium at 1 atm gave the axial alcohol almost exclusively. It is pertinent to note that in 3-alkylcyclohexanols the *cis*-isomer (both substituents equatorial) is more stable than the *trans*-isomer whereas the reverse is true for 4-alkylcyclohexanols.

Interesting results have been obtained with homogeneous catalysts: hydrogenation of 4-*t*-butylcyclohexanone in the presence of dihydridobis(triphenylphosphine)rhodium(III) ion gave approximately 86% of the equatorial (*trans*) alcohol[19], whereas with the Henbest catalyst (iridium(IV) chloride – phosphorous acid) at least 94% of the axial alcohol was obtained[20]. Catalysts prepared from di-*μ*-chlorotetrakis(cyclooctene)diiridium(I) or -dirhodium(I) and dimethyl phosphite also gave predominantly the axial alcohol[21]. A sodium chloroiridate – trimethylphosphite catalyst has been applied for the stereospecific transfer hydrogenation of steroidal ketones to the axial alcohols[22]. The stereochemical course of homogeneously catalyzed transfer hydrogenation of

2-, 3-, and 4-alkylcyclohexanones with 2-propanol as hydrogen donor was found to depend strongly on the type of catalyst. With ruthenium, rhodium, and iridium complexes the thermodynamically more stable as well as the less stable isomer could be formed preferentially[23].

Mitsui *et al.*[18] discussed their results in terms of relative rates of hydrogen transfer. It may be expected that chemisorption of cyclohexanone in a chair conformation will preferentially take place from the equatorial direction, transfer of one hydrogen atom from the catalyst surface to the carbonyl carbon atom would then give rise to a surface alcoholate with the hydrogen atom disposed equatorially. Rapid transfer of the second hydrogen atom would then lead to the axial alcohol. If, however, transfer of the second hydrogen atom is relatively slow, equilibration of the axial alcoholate with the more stable equatorial one, possibly *via* chemisorbed cyclohexanone in a boat conformation, would result in predominant formation of the equatorial alcohol. The results of heterogeneously catalyzed cyclohexanone hydrogenation seem in agreement with this rationalization: acidic conditions would certainly be expected to accelerate the transfer of the second hydrogen atom by participation of protonated solvent[24]. The effect of pressure on the rhodium catalyzed hydrogenation, very similar to the hydrogenation of 4-*t*-butylmethylenecyclohexane[25], also indicates a shift of the product controlling step. A somewhat different explanation, involving chemisorbed enol and enolate, depending on the conditions, has been presented by Augustine *et al.*[26]. In this connection it is interesting to note that deuteration of 4-*t*-butylcyclohexanone over platinum, ruthenium, iridium and osmium results essentially in the straightforward addition of deuterium. With palladium or rhodium, however, exchange takes place, showing that chemisorbed enol species are involved[27].

SELECTIVITY

Hydrogenation of alkenic aldehydes and ketones with preservation of the alkenic double bond requires special precautions — a catalyst with low activity for alkene hydrogenation has to be applied. Citral (7) has been hydrogenated selectively over platinum-iron to give geraniol[28]. Favourable results have been

Pt - FeSO₄ / EtOH

7

obtained with osmium on carbon: allyl, crotyl, and cinnamyl alcohol were formed upon hydrogenation of the corresponding aldehydes at 100° and 50-70 atm[30]; iridium has also found application as catalyst[31].

Carbocyclic arenes remain intact, the same sometimes applies to heteroaromatic systems. Thus, hydrogenation of 2-acetylpyridine (8) proceeded selectively over palladium but aselectively over rhodium or platinum, whilst 3-acetylpyridine (9) gave mixtures[32]. 4-Acetylpyridine (10) could be hydrogenated selectively over platinum[32].

Pd /C / EtOH / 3 atm

8

Pd /C / EtOH / 3 atm

9   +   40 %

73

Selective partial hydrogenation of diketones is achieved easily if the carbonyl groups differ significantly in reactivity, for example in 2,2,6-trimethylcyclohexane-1,4-dione[33] (11). Selective partial hydrogenation of symmetric diketones

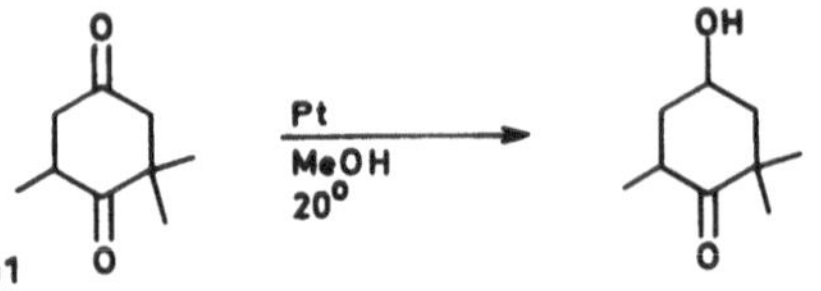

has been accomplished by interruption of the reaction after the consumption of one equivalent of hydrogen[34], which seems to indicate monopolization of the catalyst surface by diketone. The partial hydrogenation of cyclohexadeca-1,9- and cyclooctadeca-1,10-dione over platinum oxide are further examples[35].

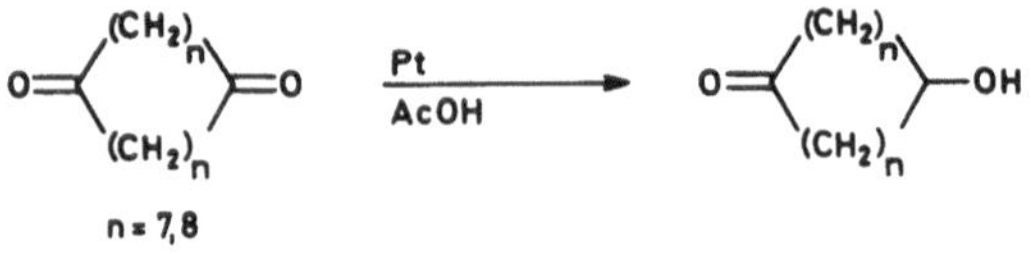

n = 7, 8

It should be noted that reduction with lithium aluminium hydride gave rise to mixtures of diketone and diol[35]. Other illustrative examples are the partial hydrogenations of aryl diketones, such as 1,4-diacetylbenzene. The selectivity is largely due to the much stronger adsorption (adsorption equilibrium constant > 7) of the diketone[34].

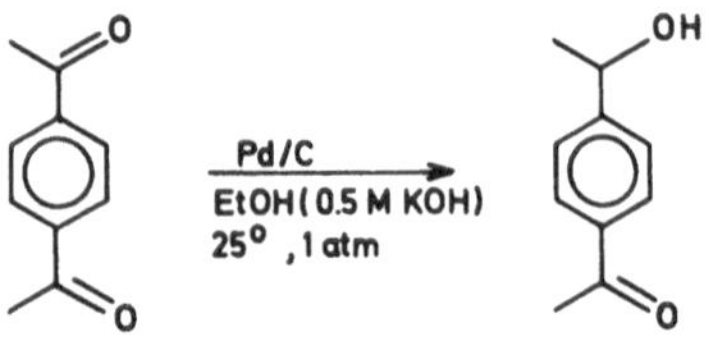

SIDE-REACTIONS

A number of side-reactions occur because of the reactivity of the carbonyl and hydroxyl functions. In particular aromatic aldehydes and ketones exhibit reductive coupling, for example vanilline[36]. Hydrogenation of 4-acetylpyridine (10) over palladium or rhodium similarly resulted in reductive coupling[32]. A

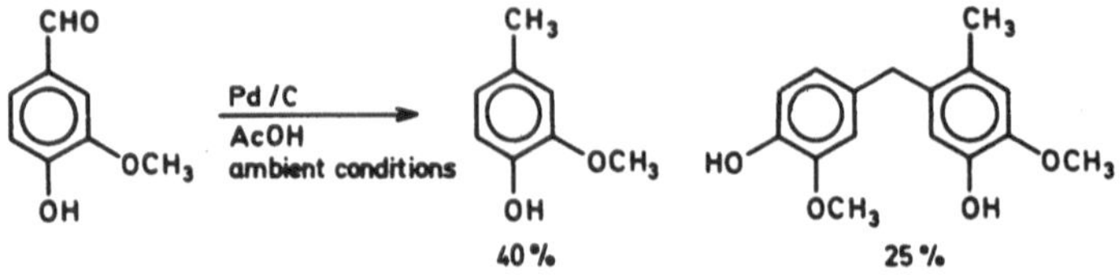

74

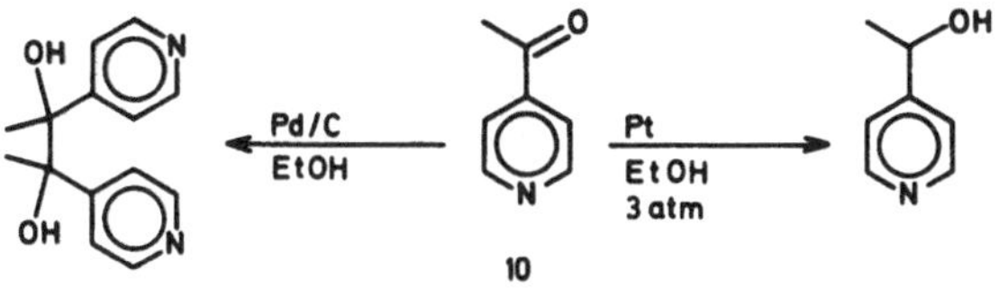

combination of hydrogenolysis and coupling occurs with 1,4- and 1,5-diketones, accompanied (in the latter case) by aldol condensation[37].

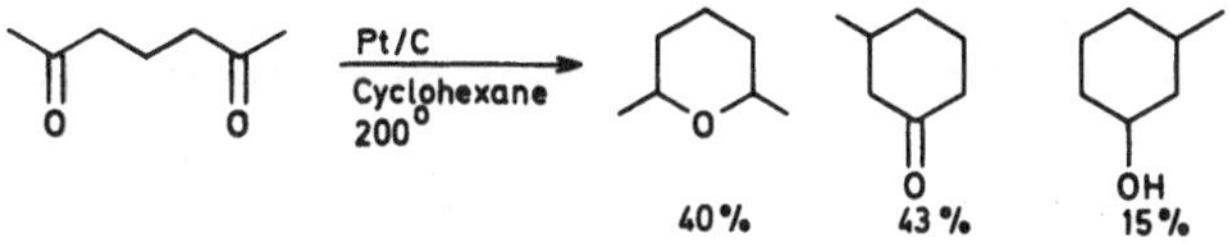

Aldolisation and coupling has also been observed with cyclic diketones[38], the cyclization of 3,3,7,7-tetrakis(ethoxycarbonyl)cycloocta-1,5-dione (12) is an interesting example[39].

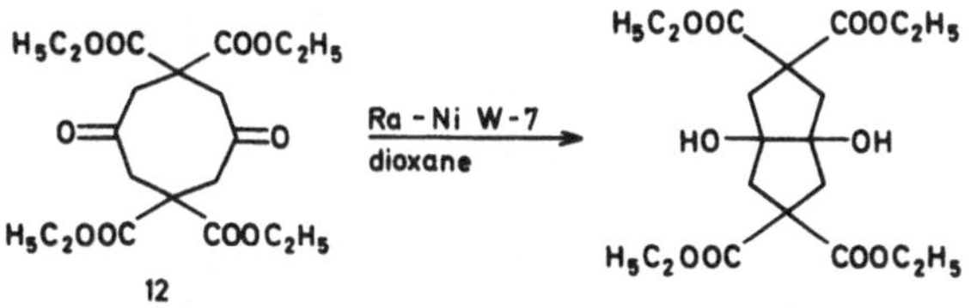

References

1. P.N. Rylander and J. Kaplan, *Engelhard Ind. Tech. Bull.*, **2**, 48 (1961).
2. W.H. Carothers and R. Adams, *J. Am. Chem. Soc.*, **45**, 1071 (1923) and **47**, 1047 (1925).
3. *Cf.* N.E. Hoffmann, A.T. Kanakkanatt, and R.F. Schneider, *J. Org. Chem.*, **27**, 2687 (1962).
4. *Cf.* R. Pietropaolo, M. Graziani, and U. Belluco, *Inorg. Chem.*, **8**, 1506 (1969); R. Pietropaolo, G. Dolcetti, M. Giustiniani, and U. Belluco, *Ibid.*, **9**, 549 (1970) and references cited therein.
5. *Cf.* ref 6.
6. V.I. Scharkov, *Chem.-Ing.-Techn.*, **35**, 494 (1963).
7. W. Wenner, *J. Org. Chem.*, **15**, 301 (1950).
8. R.I. Longley, Jr. and W. Emerson, *Org. Synth.* Coll. Vol. 4, 660 (1962).
9. R.H. Hasek, E.U. Elam, J.C. Martin, and R.G. Nations, *J. Org. Chem.*, **26**, 700 (1961).
10. G.N. Walker, *Ibid.*, **23**, 133 (1958).
11. K. Kindler, H.-G. Helling, and E. Sussner, *Justus Liebigs Ann. Chem.*, **605**, 200 (1957).

13. J.M.A. Baas, Laboratory of Organic Chemistry, Delft University of Technology, personal communication.

14. K. von Auwers, *Justus Liebigs Ann. Chem.*, **420**, 84 (1920).

15. A. Skita, *Ibid.*, **431**, 1 (1923).

16. D.H.R. Barton, *J. Chem. Soc.*, **1953**, 1027.

17. J. Klein, *Tetrahedron Lett.*, **1973**, 4307.

18. S. Mitsui, H. Saito, Y. Yamashita, M. Kaminaga, and Y. Senda, *Tetrahedron*, **29**, 1531 (1973).

19. R.R. Schrock and J.A. Osborn, *Chem. Commun.*, **1970**, 567.

20. H.B. Henbest and T.R.B. Mitchell, *J. Chem. Soc. C*, **1970**, 785.

21. M.A. Bennet and T.R.B. Mitchell, *J. Organomet. Chem.*, **70**, C30 (1974).

22. J.C. Orr, M. Mersereau, and A. Sanford, *Chem. Commun.*, **1970**, 162.

23. V.Z. Sharf, L.Kh. Freidlin, V.N. Krutii, and I.S. Shekoyan, *Izv. Acad. Nauk SSSR, Ser. Khim.*, **1974**, 1330, *Chem. Abstr.*, **81**, 177132*t* (1974).

24. F. van Rantwijk, A. van Vliet, and H. van Bekkum, *J. Chem. Soc., Chem. Commun.*, **1973**, 234.

25. S. Siegel and D.W. Ort in *Catalysis*, B. Delman and G. Jannes, Eds., Elsevier, Amsterdam, 1975, p 219.

26. R.L. Augustine, D.C. Miglorini, R.E. Foscante, C.S. Sodano, and M.J. Sisbarro, *J. Org. Chem.*, **34**, 1075 (1969).

27. Y. Takagi, S. Tetratani, and K. Tanaka, *Proc. Int. Congr. Catal. 5th, 1972*, **1**, 757 (1973).

28. R. Adams and B.S. Garvey, *J. Am. Chem. Soc.*, **48**, 477 (1926).

29. I.A. Kaye and R.S. Matthews, *J. Org. Chem.*, **29**, 1341 (1964).

30. P.N. Rylander and D.R. Steele, *Tetrahedron Lett.*, **1969**, 1579.

31. E.N. Bakhanova, A.S. Astakhova, Kh.A. Brikenshtein, V.G. Dorokhov, V.I. Savchenco, and M.L. Khidekel, *Izv. Akad. Nauk SSSR, Ser. Khim.*, **1972**, 1993, *Chem. Abstr.*, **78**, 15967*e* (1973).

32. M. Freifelder, *J. Org. Chem.*, **29**, 2895 (1964).

33. F. Hoffmann-La Roche and Co, A.-G., Brit. Patent 790,607 (1958), *Chem. Abstr.*, **52**, 15575[a] (1958).

34. H. van Bekkum, A.P.G. Kieboom, and K.J.G. van de Putte, *Rec. Trav. Chim. Pays-Bas*, **88**, 52 (1969).

35. A.T. Blomquist and J. Wolinski, *J. Am. Chem. Soc.*, **77**, 5423 (1955).

36. A. St.Phau, *Helv. Chim. Acta*, **22**, 550 (1939).

37. N.I. Shuikin and G.K. Vasilevskaya, *Izv. Akad. Nauk SSSR Ser. Khim.*, **1964**, 557, *Chem. Abstr.*, **60**, 15818[g] (1964).

38. G.L. Buchanan, J.G. Hamilton, and R.A. Raphael, *J. Chem. Soc.*, **1963**, 4606.

39. A.C. Cope and F. Kagan, *J. Am. Chem. Soc.*, **80**, 5499 (1958).

6. Hydrogenation of carbon-nitrogen multiple bonds

NITRILES

Nitriles are hydrogenated smoothly in the presence of rhodium or Raney nickel to give the corresponding primary amines; the other platinum metals and Raney cobalt can also be used. Reaction of the resulting primary amine with the imine intermediate constitutes a serious complication, as it gives rise to formation of the secondary amine. This side-reaction has been put to useful application in an industrial synthesis of piperidine derivatives[1].

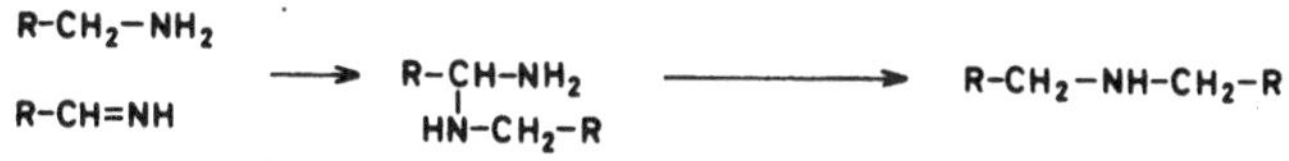

$$R-CH_2-NH_2$$
$$R-CH=NH \longrightarrow R-CH-NH_2 \quad\underset{HN-CH_2-R}{} \longrightarrow R-CH_2-NH-CH_2-R$$

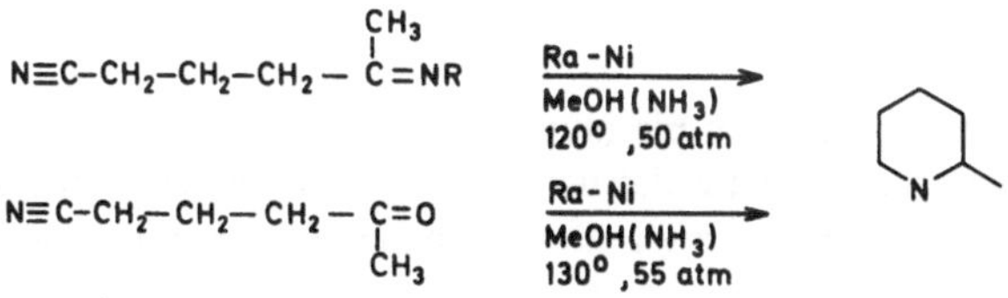

In the presence of ammonia, the formation of secondary amine is suppressed.

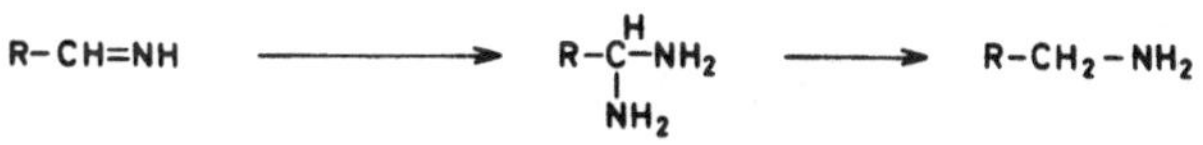

$$R-CH=NH \longrightarrow R-\overset{H}{\underset{NH_2}{C}}-NH_2 \longrightarrow R-CH_2-NH_2$$

Hydrogenation of the nitrile in liquid ammonia as solvent is a very elegant solution which has been applied to the hydrogenation of isophthalonitrile (1) over Raney cobalt[2]. Hydrogenation in alcohol/ammonia is the conventional procedure which has been applied to the hydrogenation of 3-cyanomethyl-

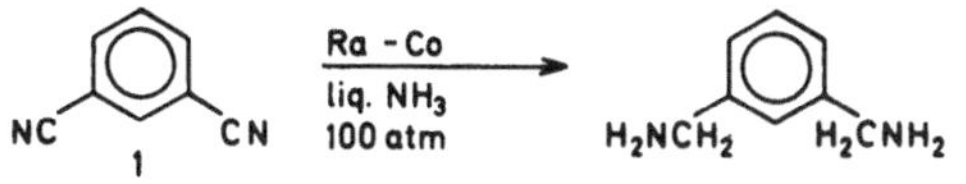

indole (2) over rhodium to give tryptamine[3]. Sodium hydroxide has been

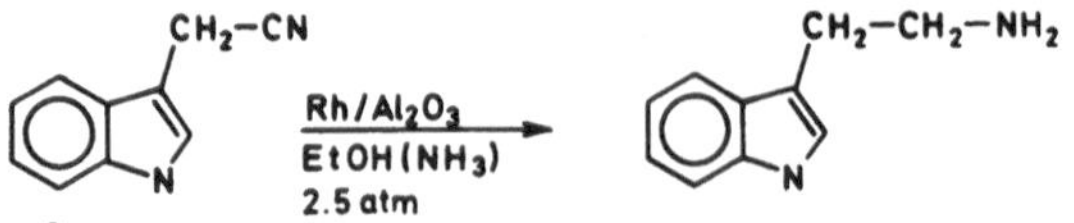

2   **78 %**

found to have an accelerating effect on the hydrogenation of 4,4'-dicyano-
bibenzyl (**3**)[4]. Scavenging of the primary amine by means of acetylation has

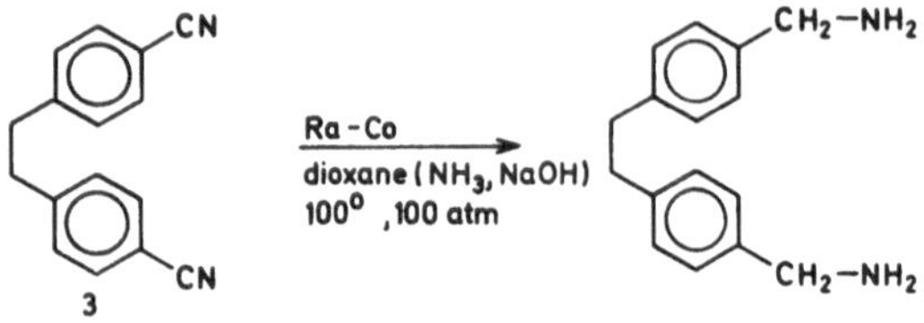

3

been applied to the hydrogenation of the photodimer of fumaronitrile (**4**) over
platinum in acetic anhydride[5].

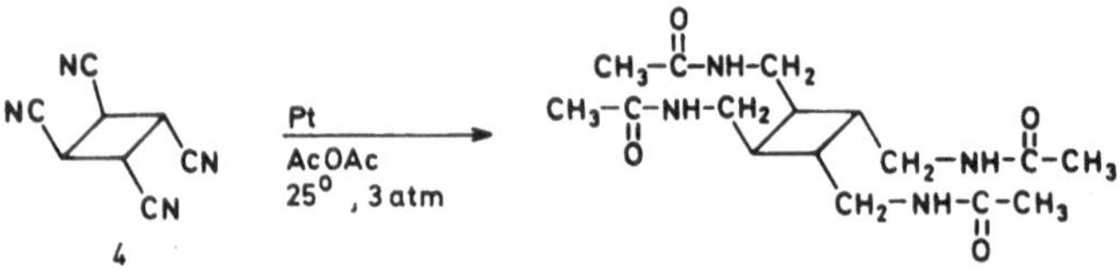

4

IMINES

Imines are usually hydrogenated over platinum or Raney nickel under ambient
conditions. Rhodium and platinum are also used as catalysts. The latter metal is
less suitable, however, for the hydrogenation of benzylideneamines, due to
subsequent hydrogenolysis of the benzyl group. An exception is found in the
hydrogenation of N-benzylidenecyclopropylamine (**5**) over palladium[6].

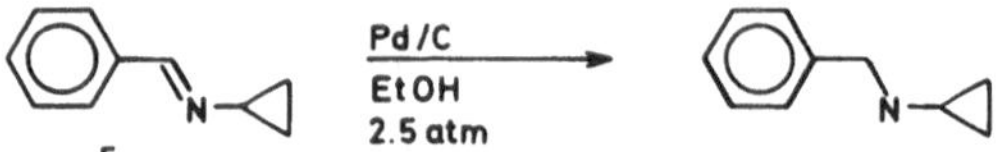

5

Although selective hydrogenation of alkenyl imines is generally impossible,
many other unsaturated functions remain preserved. Examples are the hydrogenation of N-furfurylidenecyclopropylamine[7] (**6**) and 8-aza-8,9-dehydroestron
methyl ether perchlorate[8] (**7**).

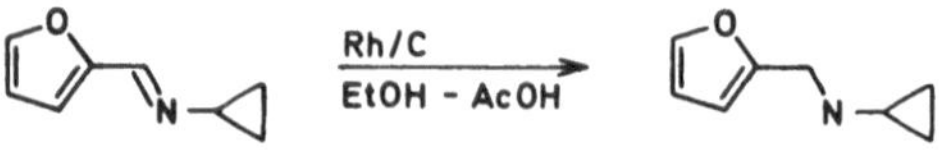

6

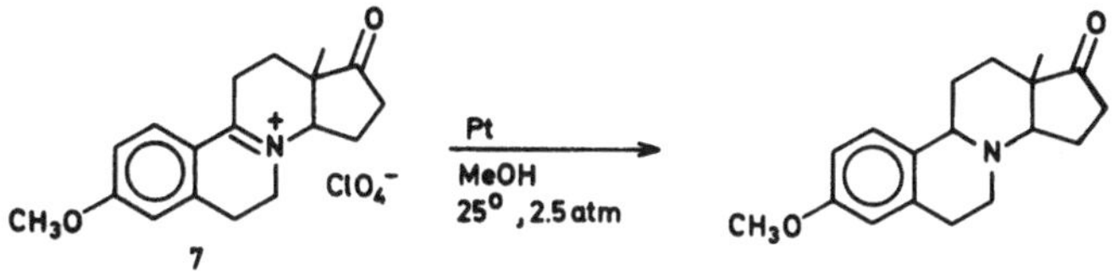

Pt
MeOH
25°, 2.5 atm

7

References

1. *Cf.* J.J.M. Deumens and J.A. Thoma, Ger. Offen. 2,014,837, *Chem. Abstr.*, **73**, 120505 (1970); S.H. Groen, J.J.M. Deumens, and P.A.M.J. Stijfs, Ger. Offen. 2,034,360, *Chem. Abstr.*, **74**, 64209 (1971); J.A. Thoma and J.J.M. Deumens, U.S. Patent 3,658,824, *Chem. Abstr.*, **77**, 61828 (1972).
2. D.V. Sokol'skii and F. Bizhanov, *Tr. Inst. Khim. Nauk, Akad. Nauk Kaz. SSSR*, **7**, 68 (1961), *Chem. Abstr.*, **57**, 9703[c] (1962).
3. M. Freifelder, *J. Am. Chem. Soc.*, **82**, 2386 (1960).
4. H. van Brederode, Thesis, Delft University of Technology, The Netherlands, 1975, p 56.
5. G.W. Griffin, J.E. Basinski, and L.I. Peterson, *J. Am. Chem. Soc.*, **84**, 1012 (1962).
6. B.W. Horrom and W.B. Martin, U.S. Patent 3,083,226, *Chem. Abstr.*, **59**, 9888[e] (1963).
7. M. Freifelder, unpublished results cited in: M. Freifelder, *Practical catalytic hydrogenation*, Wiley-Interscience, London, 1970, p 323.
8. R.E. Brown, D.M. Lustgarten, R.J. Stanaback, and R.I. Meltzer, *J. Org. Chem.*, **31**, 1489 (1966).

7. Enantioselective hydrogenation

The possibility of enantioselective hydrogenation of compounds possessing prochiral double bonds (*i.e.* asymmetric 1,1-substituted alkenes, ketones, and ketimines) has been the subject of considerable attention. Ideally, enantio-selective hydrogenation of a prochiral system should result in 100% formation of one of the enantiomeric products. The laborious process of separation of enantiomers (involving, *inter alia*, a loss of products of at least 50%, unless racemization is possible) would then be replaced by the preparation of relatively small amounts of chiral catalyst.

Enantioselective hydrogenation involves the introduction of a diastereomeric intermediate in the course of the reaction, similar to the resolution of enantiomers *via* a diastereomeric derivative. If for example, the hydrogenation of α-phenylacrylic acid is considered, it will become clear that the absolute configuration of the product is determined by the face of the molecule at which coordination and transfer of hydrogen occurs (always assuming supra-

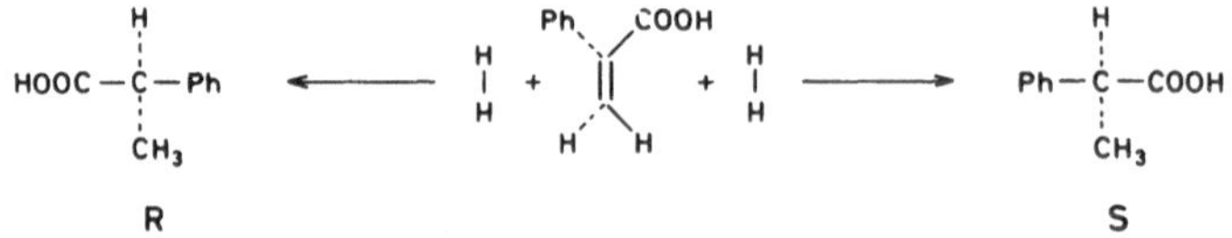

facial addition of hydrogen). In the course of the reaction, two chemisorbed states, transition states and half-hydrogenated states, exist which are mirror images. With a chiral catalyst system, these enantiomeric states become diastereomeric and may differ in energy. In order to achieve an enantioselectivity of 99%, an induced free enthalpy difference of at least 2.8 kcal mole^{-1} between the diastereomeric transition states of the rate-determining step is required.

CATALYSTS

Many attempts have been made to effect enantioselective hydrogenation over metal catalysts[1], modified by coadsorption of chiral molecules or which had been adsorbed on a chiral support, such as palladium on natural silk[2] or Raney nickel modified with L-amino acids[3]. The results have been of small practical value, although interesting mechanistic studies have resulted[4,5]. The highest enantioselectivity* was attained in the course of the hydrogenation of methyl acylacetates over modified Raney nickel (55–66%)[6].

Homogeneous chiral catalysts[7] offered the possibility of a more rational approach. Initially, little attention was paid to the molecular basis of the induction of chirality, and known active homogeneous catalysts were simply furnished with chiral ligands (usually phosphines[8,9]). On intuitive consider-ations, the chiral centre was sited as close to the metal atom as possible (i.e. on the phosphorus atom). A breakthrough was achieved when it became clear that the catalytic system as a whole (including the reactant) had to be as asymmetric and inflexible as possible. This consideration provided the basis for catalyst systems with an enantioselectivity of 70–90% in the hydrogenation of suitable alkenes.

ALKENES

In the first experiments, prochiral alkenes were hydrogenated in the presence of rhodium – chiral phosphine complexes[8-10], resulting in enantioselectivities up to 23%[8]. It soon became apparent that better results could be obtained with complexes of chirally substituted phosphines. Hydrogenation of (E)-2-phenyl-propenecarboxylic acid in the presence of chlorotris(neomenthyldiphenyl-phosphine)rhodium(I) (1) proceeded with no less than 61% enantioselectiv-

$$CH_3 —\overset{H}{\underset{CH_2COOH}{C}}— Ph$$

61% enantiomeric excess

ity[11]. In this case, the bulkiness of the chiral substituent more than compensa-ted for the distance between the chiral centre and the coordinated alkene.

The synthesis of catalyst systems with the greatest possible total steric asymmetry was carried one step further with the introduction of (4R,5R)-2,2-dimethyl-4,5-bis(diphenylphosphinomethyl)-1,3-dioxacyclopentane ('diop')[12] (2), a diphosphine based on a chiral ring system. The effect of the bidentate ligand was the suppression of the dynamic processes within the complex. Hydro-

---

* The optical purity of the product, divided by the optical purity of the catalyst

genation of amino acid precursors in the presence of rhodium-'diop' resulted, with a single exception, in 60–80% enantioselectivity[12,13]. Some results obtained with various catalysts are summarized in Table V.

Table V: Enantioselective hydrogenation of 1-phenylethenecarboxylic acid

Catalyst system[a]	Phosphine ligand	Optical yield	Ref.
Rh Cl$_3$ P$_3$/amine	methylphenylisopropylphosphine	22%	8
Rh (C$_8$H$_{12}$) P$_2^+$.BF$_4^-$	o-anisylcyclohexylmethylphosphine	16%	14
Rh Cl P$_3$/amine	neomenthyldiphenylphosphine	28%	11
Rh Cl P$_2$/amine	'diop'[b]	63%	12

a: P = phosphine ligand; C$_8$H$_{12}$ = 1,5-cyclooctadiene.
b: (4R,5R)-2,2-dimethyl-4,5-bis(diphenylphosphinomethyl)-1,3-dioxacyclopentane.

Promising results have been achieved with 'diop' variants, like **3**, **4**, and **5**. Hydrogenation of α-acetamidocinnamic acid (**6**) in the presence of rhodium complexes of these phosphines proceeded with 87.5%, 63%, and 79.5% enantioselectivity, respectively[15].

Immobilization of the chiral catalyst by attachment to a polymeric support is an obvious next step. It has been effected for the rhodium-'diop' catalyst, but led to a decrease in enantioselectivity by an order of magnitude[16].

The substituents which are present at the alkenic system may exert a crucial influence on the magnitude of the chiral induction. It was soon appreciated that free carboxylic acids like 1-phenylethenecarboxylic acid gave consistently better results than alkenes like 2-phenyl-1-butene[11]. The difference is especially striking in the case of the rhodium-'diop' catalyst: 2-phenyl-1-butene, 15% optical yield; methyl 1-phenylethenecarboxylate, 7%; the corresponding carboxylic acid, 63%[12,16]. Bifunctional interaction of the alkene (through coordination of the carboxyl group) with the metal atom, with a concomitant decrease in flexibility is the obvious explanation[11]. Enamides of type **7** have been hydrogenated in the presence of rhodium-'diop' with enantioselectivities of 45–85%. Interestingly, the sign of enantioselective hydrogenation of **7a** was

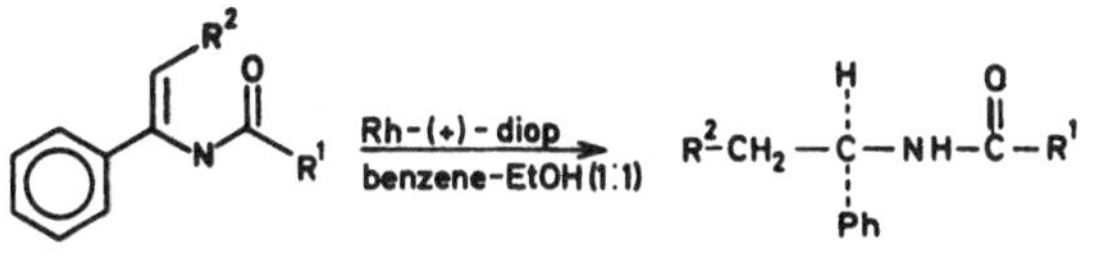

7

a $R^1 = CH_3, R^2 = H$ 45 % enantiomeric excess
b $R^1 = R^2 = CH_3$ 83 %   „
c $R^1 = C_6H_5, R^2 = CH_3$) 73%   „
d $R^1 = i\text{-}C_3H_7, R^2 = CH_3$ 85%   „

inverted when the solvent was changed from ethanol to benzene[17]. The absolute configuration of the products has been predicted in some cases by means of molecular models[18].

A chiral rhodium diphosphinite complex has been used with promising results, hydrogenation of 2-phenyl-1-butene proceeded with 33% enantioselectivity[19].

The *o*-anisylcyclohexylmethylphosphine — rhodium catalyst[14] also deserves to be mentioned. Due to the formation of a hydrogen bond between the anisyl oxygen atom and the amide function, resulting in further fixation of the reactant-catalyst system, α-acetamidoacrylic acid derivatives (**8**) were hydrogenated with up to 90% enantioselectivity[14] (see Table VI). A very similar diphosphine ligand gave even better results[20], owing to the suppression of dynamic processes in the catalyst system.

Table VI: Hydrogenation of amino acid precursors with *o*-anisylcyclohexylmethylphosphine-rhodium complex[14]

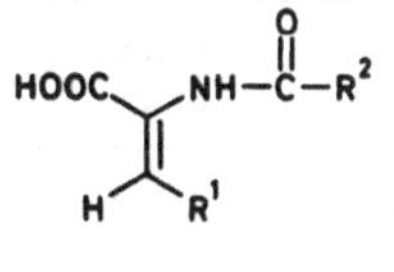

8

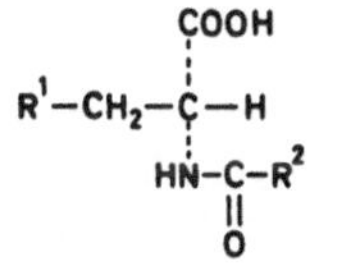

R_1	R_2	Optical yield (%)	Resulting amino acid
3-MeO-4-HO-C_6H_3	C_6H_5	90	L-DOPA
3-MeO-4-AcO-C_6H_4	CH_3	88	L-DOPA
C_6H_5	CH_3	85	L-phenylalanine
C_6H_5	C_6H_5	85	L-phenylalanine
p-Cl-C_6H_4	CH_3	77	L-*p*-chlorophenylalanine
3-(1-Ac-indolyl)	CH_3	80	L-tryptophan
H	CH_3	60	L-alanine

Examples of straightforward enantioselective hydrogenation of ketones are relatively few, reflecting the limited number of homogeneous catalysts active for ketone hydrogenation. Recently, hydrogenation of acetophenone with chiral rhodium complexes has been reported (up to 51% enantiomeric excess)[21].

A somewhat more laborious route for ketone redution, which has been applied with impressive results to enantioselective reduction, involves catalytic addition of a silane across the carbon-oxygen double bond (hydrosilation) followed by cleavage of the silyl ether. Hydrosilation of various ketones in

the presence of dihydridobis((R)-benzylmethylphenylphosphine)rhodium(III) cation with methyldiphenylsilane[22], or a similar rhodium complex and methyldiphenyl- or diethylsilane[23] gave the corresponding alcohols with enantioselectivities in the range of 32–62%. Some results are compiled in Table VII.

Table VII: Enantioselective reduction of ketones[a]

Catalyst[d]	Phosphine ligand	Silane	Aceto-phenone[a]	2-Hexa-none[a]	Ref.
RhH_2P_2	Benzylmethyl-phenylphosphine	Me_2PhSiH	32		22
$Rh_2Cl_2(C_8H_{12})_2P_4$.	idem	idem	43	3	23
idem	idem	Et_2SiH_2		30	23
$RhClP_2$	'diop'	idem	25[b]		16
idem	idem	$Ph(C_{10}H_7)SiH_2$ [c]	53[b]		16

a: enantioselectivity in %.
b: immobilized catalyst gave identical results.
c: $C_{10}H_7$ = 1-naphthyl.
d: P = phosphine ligand; C_8H_{12} = 1,5-cyclooctadiene.

The available results[16,23] indicate that ketone and silane have to be matched in order to obtain optimal results. Propyl pyruvate and ethyl phenylglyoxylate have been reduced via hydrosylation with enantioselectivities up to 80%; in this case the rhodium-'diop' catalyst gave superior results[24].

α,β-Unsaturated ketones may undergo 1,4-addition under hydrosilation conditions, with formation of the alkenyl silyl ether which yields the saturated ketone upon hydrolysis. Thus, enantioselective synthesis of chiral ketones becomes possible, although as yet in low optical yields[25].

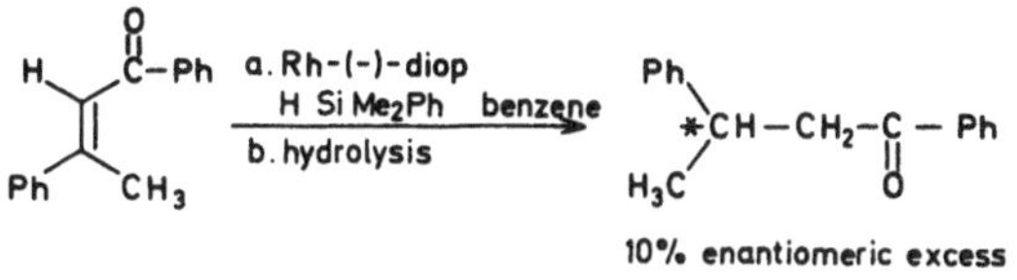

a. Rh-(-)-diop
H SiMe$_2$Ph   benzene
b. hydrolysis

10% enantiomeric excess

IMINES

The hydrosilation route has also been applied to the enantioselective reduction of imines. Thus, hydrosilation of **9** with diphenylsilane in the presence of rhodium-'diop' gave, upon hydrolysis, benzyl(1-phenylethyl)amine in 50–65% optical yield[17]. Similarly, **10** gave 1,2,3,4-tetrahydropapaverine with 39% enantioselectivity[17].

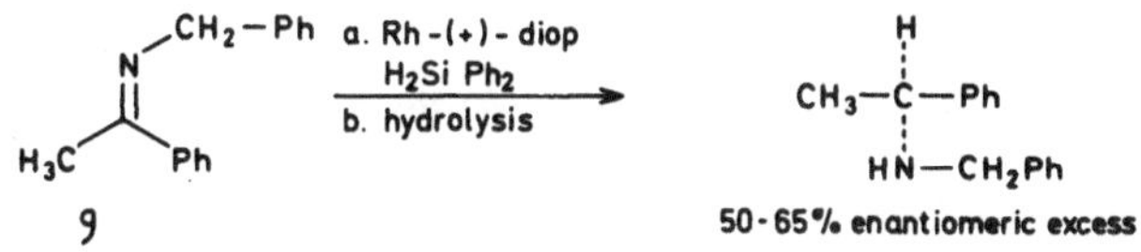

a. Rh-(+)- diop
H$_2$Si Ph$_2$
b. hydrolysis

50-65% enantiomeric excess

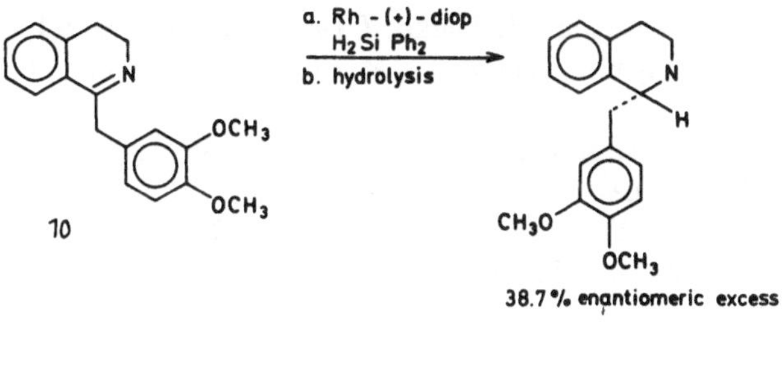

a. Rh - (+) - diop
H$_2$Si Ph$_2$
b. hydrolysis

38.7% enantiomeric excess

References

1. For a review see: E.I. Klabunovskii and E.S. Letvina, *Russ. Chem. Rev.* (Engl. Transl.), **39**, 1035 (1970); Y. Izumi, *Angew. Chem.*, **83**, 956 (1971).
2. S. Akabori, Y. Izumi, Y. Fujii, and S. Sakurai, *Nature*, **178**, 323 (1956).
3. See pertinent literature cited in ref 1.
4. I. Yasumori, Y. Inone, and K. Onabe in *Catalysis*, B. Delman and G. Jannes, Eds., Elsevier, Amsterdam, 1975, p 41.

85

5. J.A. Groenewegen and W.M.H. Sachtler, *J. Catal.*, **38**, 501 (1975).
6. T. Tanabe and Y. Izumi, *Bull. Chem. Soc. Jpn.*, **46**, 1550 (1973); L.H. Gross and P. Rys, *J. Org. Chem.*, **39**, 2429 (1974).
7. For a review see: L. Markó and B. Heil, *Catal. Rev.*, **8**, 269 (1973); H.B. Kagan, *Pure Appl. Chem.*, **43**, 401 (1975); J.D. Morrison, W.F. Masler, and M.K. Neuberg, *Adv. Catal.*, **25**, 81 (1976).
8. W.S. Knowles and M.J. Sabacky, *Chem. Commun.*, **1968**, 1445.
9. L. Horner, H. Siegel, and H. Büthe, *Angew. Chem.*, **80**, 1034 (1968).
10. L. Horner and H. Siegel, *Phosphorus*, **1**, 209 (1972).
11. J.D. Morrison, R.E. Burnett, A.M. Aguiar, C.J. Morrow, and C. Phillips, *J. Am. Chem. Soc.*, **93**, 1301 (1971); R.E. Burnett, Thesis, University of New Hampshire, 1971, *Diss. Abstr. Int. B*, **32**, 3842 (1972).
12. H.B. Kagan and T.P. Dang, *J. Am. Chem. Soc.*, **94**, 6429 (1972).
13. G. Gelbard, H.B. Kagan, and R. Stern, *Tetrahedron*, **32**, 233 (1976).
14. W.S. Knowles, M.J. Sabacky, and B.D. Vineyard, *Ann. N.Y. Acad. Sci.*, **214**, 119 (1973) and *J. Chem. Soc., Chem. Commun.* **1972**, 10.
15. T.P. Dang, J.-C. Poulin, and H.B. Kagan, *J. Organomet. Chem.*, **91**, 105 (1975).
16. W. Dumont, J.-C. Poulin, T.P. Dang, and H.B. Kagan, *J. Am. Chem. Soc.*, **95**, 8295 (1973).
17. H.B. Kagan, N. Langlois, and T.P. Dang, *J. Organomet. Chem.*, **90**, 353 (1975).
18. R. Glaser, *Tetrahedron Lett.*, **1975**, 2127.
19. M. Tanaka and I. Ogata, *J. Chem. Soc., Chem. Commun.*, **1975**, 735.
20. W.S. Knowles, M.J. Sabacky, B.D. Vineyard, and D.J. Weinkauff, *J. Am. Chem. Soc.*, **97**, 2567 (1975).
21. B. Heil, S. Torös, S. Vastag, and L. Markó, *J. Organomet. Chem.*, **94**, C47 (1975). See also: T. Hayashi, T. Mise, and M. Kumada, *Tetrahedron Lett.*, **1976**, 4351.
22. K. Yamamoto, T. Hayashi, and M. Kumeda, *Ibid.*, **54**, C45 (1973).
23. I. Ojima and T. Kogure, *Chem. Lett.*, **1973**, 541.
24. I. Ojima, T. Kogure, and Y. Nagai, *Tetrahedron Lett.*, **1974**, 1889.
25. T. Hayashi, K. Yamamoto, and M. Kumada, *Ibid.*, **1975**, 3.

III. Hydrogenolysis

1. Introduction

The hydrogenolysis reaction is defined as the reductive cleavage of a σ-bond according to

$$A-B \;+\; 2\,H \longrightarrow A-H \;+\; H-B$$
$$\quad\quad\quad\quad |$$
$$\quad\quad\quad\quad *$$

in which A and B may be H, C, O, N, S, and the halogens. The reaction may be considered as a heterogeneously or homogeneously catalyzed displacement reaction with hydrogen (atomic hydrogen or hydride) as the attacking species.

CHEMISORPTION

In contrast to the hydrogenation reaction, which involves the reductive cleavage of π-bonds, homogeneously catalyzed hydrogenolysis has been observed in just a few cases (see below). In the case of heterogeneously catalyzed hydrogenolysis, the A-B σ-bonds show little affinity to the catalyst surface. Therefore good catalytic activity of the transition metals is only observed if A or B itself shows sufficient bond formation with the metal (*e.g.* sulphur, the halogens, and the cyclopropane ring because of the enhanced *p*-character of the C-C bonds) or when the A-B bond is situated in the neighbourhood of an unsaturated function such as an aryl, vinyl or carbonyl group. The unsaturated function serves as a 'handle' that brings the A-B bond closer to the catalytic surface, thus promoting overlap between the σ and σ^* orbitals of A-B and the *d* and *spd* orbitals of the transition metal. In addition, the hydrogenolysis of the A-B bond will be generally activated by the electronic effects in the 'handle'.

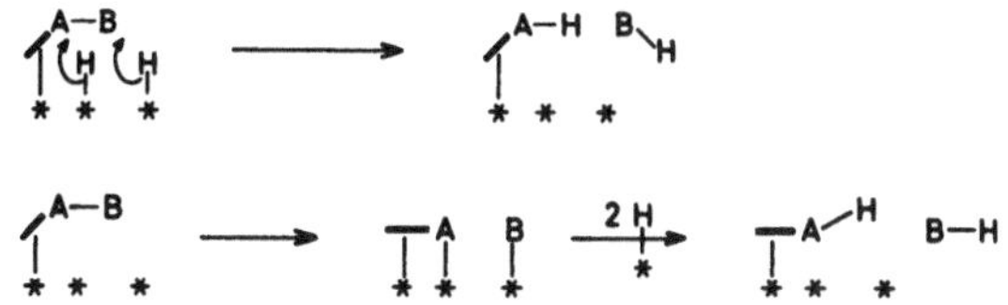

As a consequence, the strength of adsorption of the 'handle-containing' reaction product A-H will often be comparable to that of the reactant A-B, *i.e.* the product formed will compete with the reactant for active sites on the catalyst. Thus, most hydrogenolyses proceed at a progressively decreasing rate of reaction because the adsorption equilibrium causes a decreasing surface concentration of A-B during the reaction.

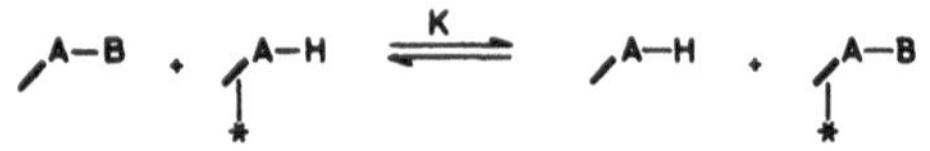

KINETICS

Let us assume for simplicity (which proves to be the case in most liquid-phase hydrogenolyses) that no real competition between hydrogen and the reactant occurs on the catalytic surface. The reaction rate is then given by

$$-\frac{d[AB]}{dt} = r = k.\,\Theta_{AB}.w \tag{1}$$

with

$$\Theta_{AB} = \frac{b_{AB}[AB]}{1 + b_{AB}[AB] + \Sigma bc} \tag{2}$$

according to Langmuir adsorption. k is the reaction rate constant, which contains the hydrogen concentration on the catalyst, Θ_{AB} is the fraction of the active catalyst surface covered by A-B, b_{AB} is the adsorption constant for A-B, Σbc is the sum of the contributions of the solvent and the hydrogenolyzed products to the denominator of the Langmuir expression, and w the amount of catalyst. Combination of (1) and (2) gives

$$r = k.\,\frac{b_{AB}[AB]}{1 + b_{AB}[AB] + \Sigma bc}.\,w \tag{3}$$

If both A-B and A-H are strongly adsorbed on the catalyst, *i.e.*

$b_{AB}[AB] + b_{AH}[AH] \gg 1 + \Sigma bc$, we may write

$$r = k.\,\frac{b_{AB}[AB]}{b_{AB}[AB] + b_{AH}[AH]}.\,w \tag{4}$$

With $\dfrac{b_{AB}}{b_{AH}} = K$, this relation becomes

$$r = k.\,\frac{K}{K + \dfrac{[AH]}{[AB]}}.\,w \tag{5}$$

In this way, the quantitative influence of K as well as the conversion on the rate of reaction may be conveniently evaluated. Lower rates of reaction will occur at high conversions (high [AH]/[AB] ratios), whereas reactions for which K is small will proceed with a drastic declining rate. In some instances, the reactant A-B is much more strongly adsorbed than both reaction products A-H and B-H, *e.g.*

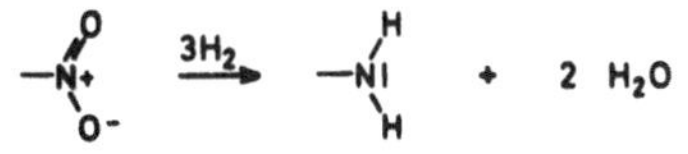

Here, $K \gg 1$ and the conversion proceeds almost linearly with the time of reaction ($r = k$). In other words, the catalyst is completely covered by A-B during the reaction, *i.e.* $\Theta_{AB} \approx 1$ in relation (1). Competition between the solvent and the reactant A-B for active sites on the catalyst surface often causes a slight progressive decrease of reaction rate, in particular at higher conversions.

SELECTIVITY

If simultaneous hydrogenolysis of two σ-bonds is possible, *e.g.*

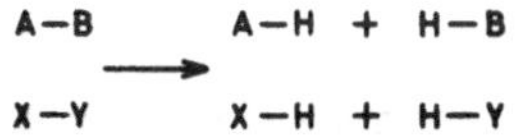

$$
\begin{array}{lll}
A-B & A-H & + \; H-B \\
\longrightarrow & & \\
X-Y & X-H & + \; H-Y
\end{array}
$$

or

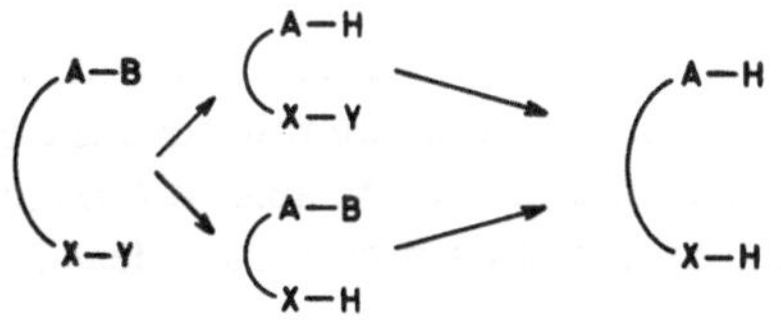

the selectivity of the reaction depends on both the relative reaction constant and the relative adsorption constant of the A-B and X-Y bonds. Quantitatively, the selectivity (S) may be expressed by

$$
S = \frac{r_{AB}}{r_{XY}} = \frac{b_{AB} \cdot k_{AB} \cdot [AB]}{b_{XY} \cdot k_{XY} \cdot [XY]}
$$

which follows directly from the rate equation (3) simplified on page 90. A similar case is presented by

In the case of two (or more) hydrogenolyzable σ-bonds in one and the same molecule (the second example given above), the selectivity may further be strongly dependent on the geometry of the reactant. For instance, if A-B and C-D possess comparable reaction rate constants whereas A-B is more strongly adsorbed than C-D, a much higher selectivity towards A-B has to be expected when A-B and C-D cannot become chemisorbed on the catalyst concomitantly

as compared to the case in which the adsorption of A-B also brings the C-D bond in close contact with the catalyst surface:

In the latter case the selectivity, which is partly due to the difference in strength of adsorption, can be lost to a large extent.

CATALYSTS

The catalysts of choice for the hydrogenolysis reaction are almost exclusively palladium and Raney nickel. Platinum and rhodium have been applied in some special cases; whereas ruthenium shows the lowest hydrogenolytic activity. The remarkable activity of palladium towards the hydrogenolysis reaction is probably due to its homolytic/heterolytic ambivalent nature towards the cleavage of the H-H bond of hydrogen. Palladium seems to be able to produce a reducing agent with hydride as well as atomic hydrogen character, dependent on the substrate and reaction medium employed.

In general, homogeneous catalysts exhibit no or very little activity for the hydrogenolysis reaction. Firstly, the A-B σ-bond itself will not be able to achieve a sufficient degree of bond formation with the metal atom of the homogeneous catalyst, as noted before for metallic catalysts. Secondly, an unsaturated 'handle' at A-B will block the active site (vacant position) of the catalyst and thus prevents any further interaction between the A-B σ-bond and the metal atom. It seems that the hydrogenolysis reaction preferably requires at least two (neighbouring) active sites. This condition, of course, is more easily fulfilled for the heterogenous catalysts. Nevertheless, homogeneous catalysts may, *in principle*, have activity towards the hydrogenolysis reaction if exchange occurs between one of the ligands and the 'handle' of the reactant:

92

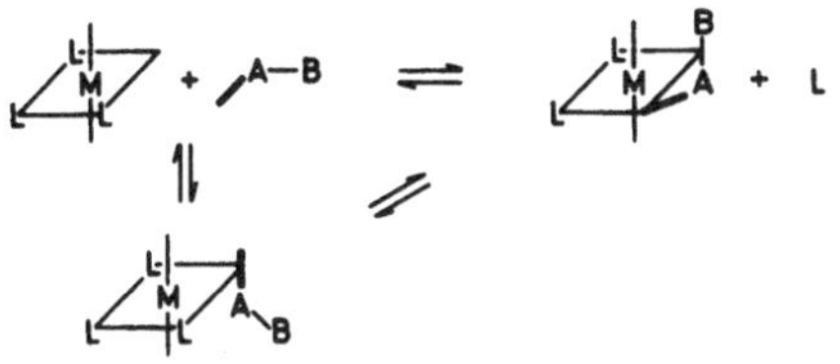

Indeed, a few hydrogenolytic active homogeneous catalysts have been recently described.

MECHANISM

The mechanism of the hydrogenolysis reaction may be often described in terms of homolytic or heterolytic displacement reactions, whereby A or B of the A-B σ-bond is considered as the leaving group. In this way, various phenomena concerning stereoselectivity and rate of the reaction are conveniently explained by the use of general concepts derived from other organic reactions. Furthermore, the reaction pathway (including the nature of adsorbed organic species in the initial and transition state and the character of the reducing agent) closely resembles that of the hydrogenation reaction in most instances. The hydrogenolysis reaction of the C-H, C-C, C-O, C-N, C-S, C-Hal and some hetero-hetero bonds are dealt with in sequence in the paragraphs that follow and are treated from both a mechanistic and a preparative point of view.

2. Carbon-hydrogen hydrogenolysis (exchange)

In most instances, hydrogenation and hydrogenolysis reactions will be accompanied by hydrogenolysis of carbon-hydrogen bonds according to

This reaction is just observed when using isotopes of hydrogen as the reducing agent, and is commonly denoted as H/D and H/T exchange. As an example, deuterolysis of 2-phenyl-2-propanol to cumene was accompanied by exchange of the β-hydrogens due to the close contact of the α-methyl groups with the catalyst surface during the cleavage of the carbon-oxygen bond:

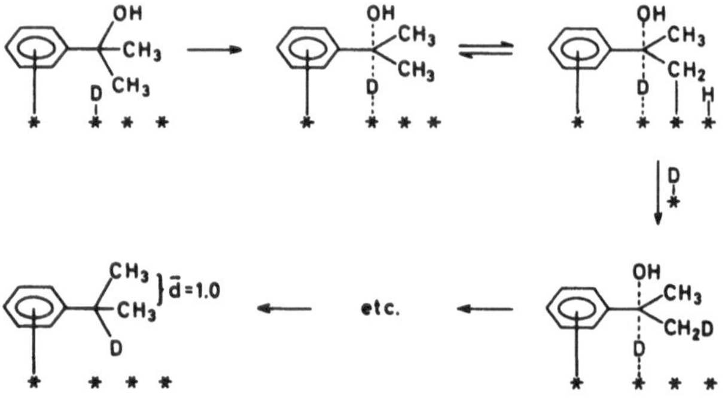

resulting in an average incorporation of one deuterium atom at the α-methyl groups[1].

During deuterolysis reactions, for instance, isotopic dilution will occur because of occupation of a part of the catalyst surface by the exchanged hydrogen atoms. For this reason, deuteric solvents like deuterium oxide, alcohols-O-d and acids-O-d are most often applied as the reaction medium. In this way, the relatively fast exchange between the active deuterium of the solvent and the chemisorbed hydrogen[2] according

$$\underset{*}{\overset{|}{H}} + \underset{*}{\overset{|}{D^{+}}} \rightleftharpoons \underset{*}{\overset{|}{D}} + H^{+}$$

to a large extent cleans the metal surface of the hydrogen from the deuterolysis reaction. As a rule, the chemisorbed hydrogen (or deuterium) already present on the catalyst surface is sufficient to initiate the exchange reaction so that no deuterium gas has to be supplied when using one of the above-mentioned deuteric solvents.

The deuterolysis of the carbon-hydrogen bond has been much studied from a mechanistic point of view and with the aim of the development of suitable techniques for the preparation of deuterated compounds.

(sp^3)CARBON-HYDROGEN BONDS

Deuterolysis of the carbon-hydrogen bonds of alkanes usually requires very active metallic catalysts or rather drastic reaction conditions. The absence of unsaturated functional groups prohibits good non-dissociative chemisorption of the alkane on the catalyst surface, and results in a less efficient catalytic support with respect to the carbon-hydrogen cleavage as compared with compounds bearing unsaturated functions. The rate of C-H dissociation slows in the order *tert* C-H > *sec* C-H > *prim* C-H.

Extensive studies[3-7] with several model alkanes and cycloalkanes reveal the preferential formation of eclipsed 1,2-diadsorbed alkanes (or π-adsorbed alkenes) by the loss of two hydrogen atoms at adjacent carbon atoms. For example, norbornane yields selectively *exo*-2,3-dideuteronorbornane (1) as the *initial* reaction product[5].

(In all instances, the reaction mixture has been analyzed at very low conversions, *i.e.* each product has been formed during just one residence on the catalyst surface).

Deuteration of norbornene gives the identical dideutero product. This points to the occurence of similar species for both the carbon-carbon double bond hydrogenation (*cf.* p 4) and carbon-hydrogen hydrogenolysis. Staggered 1,2-diadsorbed alkanes are not formed, or are only formed with difficulty, as appears from the selective formation of adamantane-d_1 as the initial product of deuterolysis of adamantane.

95

Mono-adsorbed alkane species are formed less readily than the eclipsed 1,2-diadsorbec species and are hardly converted to 1,4-diadsorbed alkanes exemplified by

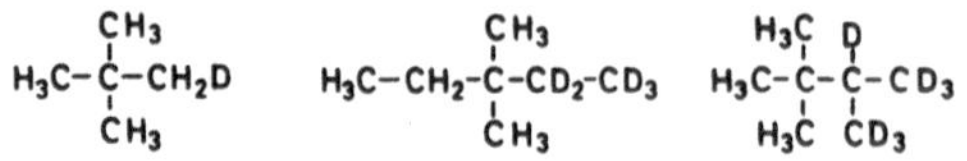

$$H_3C-\overset{\overset{\displaystyle CH_3}{|}}{\underset{\underset{\displaystyle CH_3}{|}}{C}}-CH_2D \qquad H_3C-CH_2-\overset{\overset{\displaystyle CH_3}{|}}{\underset{\underset{\displaystyle CH_3}{|}}{C}}-CD_2-CD_3 \qquad H_3C-\overset{\overset{\displaystyle H_3C}{|}}{\underset{\underset{\displaystyle H_3C}{|}}{C}}-\overset{\overset{\displaystyle D}{|}}{\underset{\underset{\displaystyle CD_3}{|}}{C}}-CD_3$$

as the highest exchanged products from the deuterolysis reaction of the corresponding hydrogen compounds. For monocyclic alkanes one would expect that, during one residence of the molecule on the catalyst surface, only hydrogen atoms at the same side of the ring would be exchanged by deuterium, as depicted below for cyclopentane.

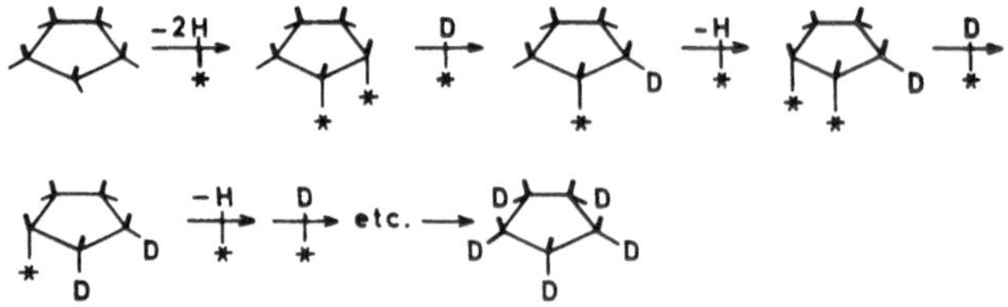

Continued alteration between mono- and 1,2-diadsorbed species would result in the incorporation of five deuterium atoms. However, the initial product contained appreciable amounts of cyclopentane-d_8 and -d_{10}. At first intermediate α,α-diadsorbed (carbene) species were assumed, e.g.

which could explain two-side exchange. Alternatively, the data have been interpreted by the so-called π-allyl mechanism[8-10]. The formation of a π-adsorbed intermediate was postulated*, involving the loss of three hydrogen atoms at three consecutive carbon atoms, followed by a top-side attack of a physically adsorbed deuterium molecule:

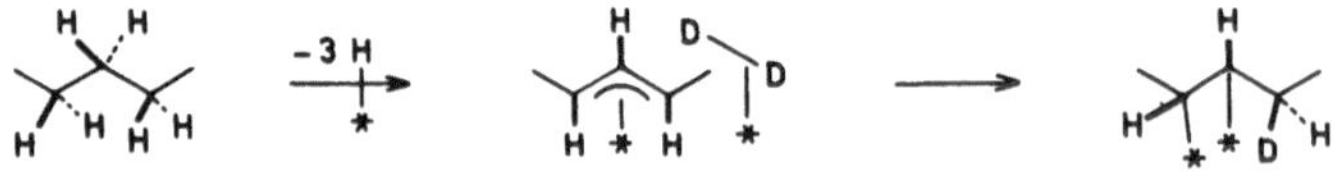

Subsequently, a more plausible explanation, however, has been given by the roll-over mechanism[3-7]. Interconversion of 1,2-diadsorbed alkanes is considered to occur via a 1,2-tetraadsorbed alkene species 2 as formulated below for cyclopentane.

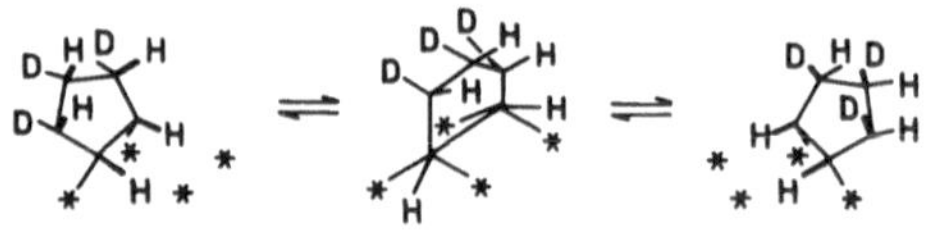

2

One interconversion gives cyclopentane-d_8, whereas at least two interconversions are necessary to explain the formation of cyclopentane-d_{10} during just one residence on the catalyst surface.

---

* For a review concerning the structure and relative stability of alkane intermediates on metallic catalysts see ref [11].

96

It may be noted that this mechanism permits exchange to maximal d_5-products for an isolated trimethylene unit:

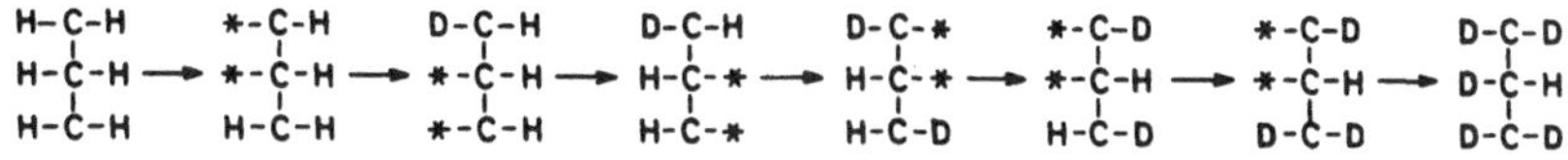

```
H-C-H      *-C-H      D-C-H      D-C-H      D-C-*      *-C-D      *-C-D      D-C-D
  |          |          |          |          |          |          |          |
H-C-H  →   *-C-H  →   *-C-H  →   H-C-*  →   H-C-*  →   *-C-H  →   *-C-H  →   D-C-H
  |          |          |          |          |          |          |          |
H-C-H      H-C-H      *-C-H      H-C-*      H-C-D      H-C-D      D-C-D      D-C-D
```

This, indeed, was observed for 1,1,3,3-tetramethylcyclohexane, giving the d_5-product[5,9] (in addition to the d_3 product from exchange at one side of the cyclohexane ring) which finally excluded roll-over of the alkane species *via* a 1,1-diadsorbed intermediate (3),

since then all hydrogens of the trimethylene unit should be able to be exchanged.

Saturated hydrocarbons will yield most often a complex mixture of deuterated products, in particular at high conversions, because of repeated dissociative adsorption of the products initially formed during the deuterolysis reaction.

Relatively selective H/D exchange may occur if unsaturated groups are present in the molecule as may be seen from the deuterolysis of a number of alkylbenzenes over nickel on kieselguhr in deuterium oxide at $100°$[12]:

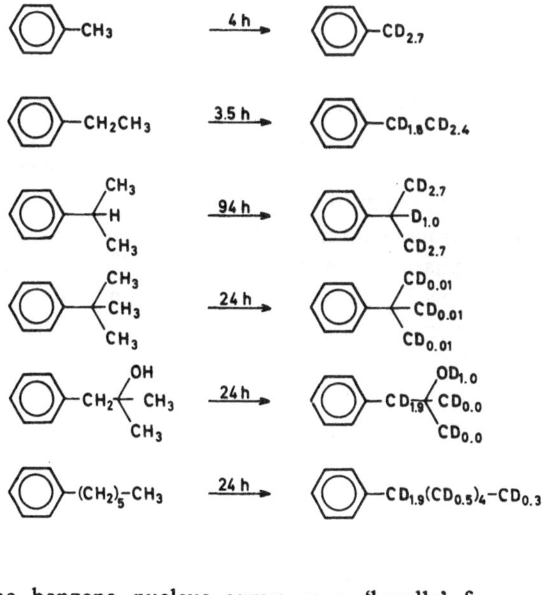

The benzene nucleus serves as a 'handle' for good chemisorption of the molecule on the catalyst surface and particularly promotes the hydrogenolysis of α- and β-carbon-hydrogen bonds of the alkyl chain by the formation of benzylic (4) and styrene-like (5) adsorbed species.

4   5

The homogeneously platinum catalyzed deuterolysis of alkylbenzenes takes place at both the aliphatic and aromatic C-H bonds. With the exception of toluene, alkyl hydrogen exchange is slow relative to aromatic hydrogen exchange[13]. Homogeneous platinum(II) catalyzed H/D exchange at the alkyl part of 1,1-dimethylpropylbenzene[14] and alkenes of the type $RCH_2C(CH_3)_2CH = CH_2$[15] takes place selectively at the γ-position with respect to the unsaturated function (α-exchange cannot occur in these compounds). Co-ordination of the aromatic ring or the olefinic double bond to the platinum brings the alkyl group into close proximity to the metal. Subsequent insertion of the metal at the γ-carbon-hydrogen bond gives rise to five-membered ring intermediates 6 and 7 which are responsible for the regiospecific manner of the H/D exchange.

6   7

Almost complete H/D exchange of saturated fatty acid esters may be accomplished over platinum metal in deuterium oxide (up to 200° for several days). Here, the ester function will act as the 'handle' and thus promotes the deuterolysis of the carbon-hydrogen bonds of the aliphatic chain.

Surprisingly, platinum(II) and iridium(III) chloro complexes have been found to catalyze deuterolysis of aliphatic carbon-hydrogen bonds[16,17]. Extensive exchange (25–91%) can be achieved with $PtCl_4{}^{2-}$ in deuterium oxide-acetic acid-O-d at 100–120° and prolonged reaction times (up to 137 h). Here, the rate of reaction decreases in the order: *prim* C-H > *sec* C-H > *tert* C-H. The rate increases with increase in the chain length of the alkane, decreases with increase in branching, and is highest for the cyclic alkanes.

Finally, very selective and rapid deuterolysis of the hydrogens at the C-1 position in primary aliphatic alcohols takes place with dichlorotris(triphenyl-phosphine)ruthenium(II) in deuterium oxide at 200°[18].

$$CH_3(CH_2)_8CH_2OH \xrightarrow[\substack{D_2O \\ 200°, \ 0.5 \ h}]{(Ph_3P)_3RuCl_2} CH_3(CH_2)_8CD_{1.9}OD$$ 93%

Other transition metal complexes (Rh, Pt, Pd, Ir, Ni) exhibit no detectable catalytic activity, whereas in the case of *secondary* alcohols there is a small amount of exchange in addition to extensive dehydration. Probably, the reaction proceeds via the following hexacoordinated species:

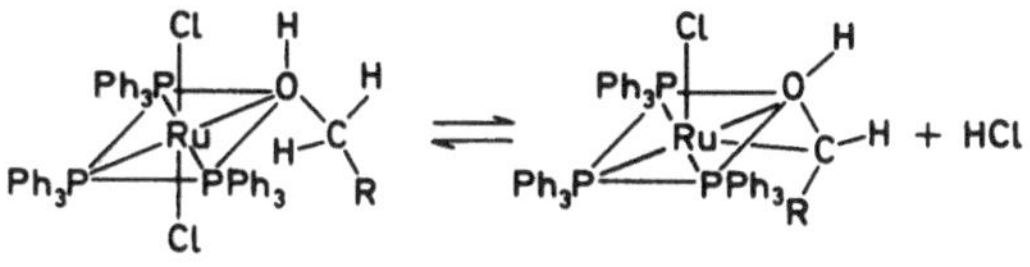

in combination with the rapid equilibrium:

$$HCl + D_2O \rightleftharpoons DCl + HDO$$

(sp^2)CARBON-HYDROGEN BONDS

Heterogeneously catalyzed H/D exchange of aromatic hydrogen atoms occurs less readily than the 'activated' aliphatic hydrogen atoms such as the α-benzylic or α-allylic hydrogens. In particular, the rate of H/D exchange is decreased when steric repulsions are involved between substituents in the aromatic nucleus and the catalyst surface; this is reflected in the rather slow hydrogenolysis of carbon-hydrogen bonds in the *ortho*-position to a bulky substituent[19]. No electronic effects of the substituents on the rate of exchange of aromatic hydrogens have been observed over platinum in the liquid phase. On the other hand, the rates of exchange for a number of mono-substituted benzenes in the gas phase over palladium were found to be clearly dependent on the electronic nature of the substituent and pointed to a nucleophilic substitution reaction with some negative charge development on the aromatic nucleus according to[20]

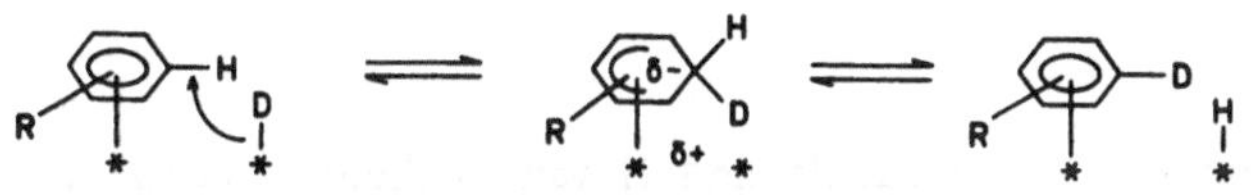

As a rule, the hydrogenolysis of aromatic carbon-hydrogen bonds is described in terms of dissociative (the mechanism generally accepted for (sp^3)carbon-hydrogen hydrogenolysis) as well as associative reaction mechanisms[21] which may be depicted as

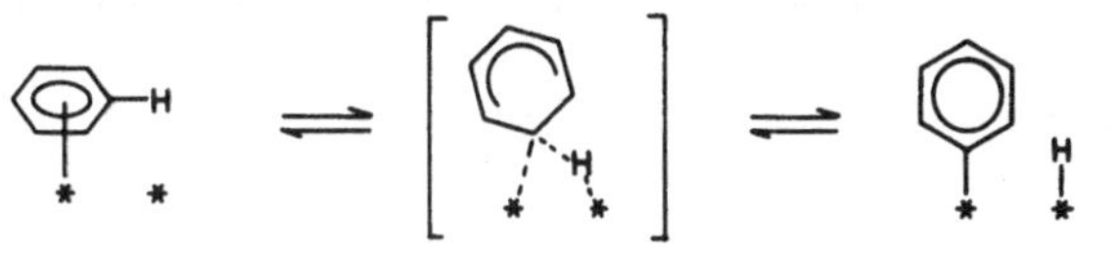

and

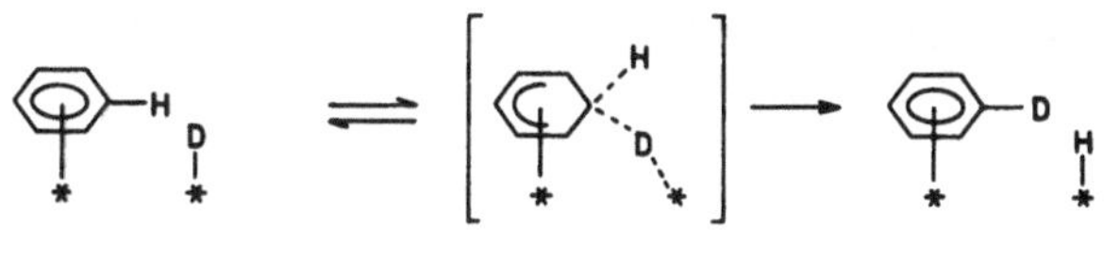

99

respectively[22]. Up to the present, the phenomena observed are in agreement with both types of mechanism and do not enable a choice between them to be made.

A preparative method for extensive hydrogen-deuterium exchange for a number of aromatic and heteroaromatic compounds has been described using platinum as the catalyst under rather drastic conditions (up to $250°$)[23].

$$d_5 = 9.4\,\%$$
$$d_6 = 8.4\,\%$$
$$d_7 = 82.2\,\%$$

Aromatic hydrogens may also be exchanged using *homogeneous* catalysts[13,16,17,24-27] either with the aid of molecular deuterium (using iridium and rhenium complexes) or deuteric solvents like deuterium oxide, alcohols-*O*-d and acids-*O*-d (using chloroplatinum(II) complexes). Very rapid hydrogen-deuterium exchange occurs according to

$$C_6D_6 + \text{arene } H_m \underset{\substack{25° \\ \text{few min}}}{\rightleftarrows} C_6D_{(6-n)}H_n + \text{arene } H_{(m-n)}D_n$$

with Friedel-Crafts catalysts ($AlCl_3$, $SbCl_5$, BBr_3, WCl_6, $TaCl_5$)[28]. No exchange of hydrogens of aliphatic side chains could be detected. The use of a large excess of hexadeuterobenzene affords the perdeutero derivatives of other arenes in excellent yield.

By analogy with the above-mentioned heterogeneously catalyzed deuterolysis of aromatic C-H bonds, the phenomena for the homogeneously catalyzed reaction have been explained by both associative and dissociative mechanisms[21].

References

1. A.P.G. Kieboom, J.F. de Kreuk, and H. van Bekkum, *J. Catal.*, **20**, 58 (1971).
2. J.J. Philipson and R.L. Burwell, Jr., *J. Am. Chem. Soc.*, **92**, 6125 (1970).
3. R.L. Burwell, Jr., and K. Schrage, *Discuss. Faraday Soc.*, **41**, 215 (1966).
4. J.A. Roth, B. Geller, and R.L. Burwell, Jr., *J. Res. Inst. Catal. Hokkaido Univ.*, **16**, 221 (1968).
5. R.L. Burwell, Jr., *Acc. Chem. Res.*, **2**, 289 (1969).
6. R.L. Burwell, Jr., *Catal. Rev.*, **7**, 25 (1972).
7. H.A. Quinn, J.H. Graham, M.A. McKervey, and J.J. Rooney, *J. Catal.*, **26**, 326 (1972), and references.
8. F.C. Gault, J.J. Rooney, and C. Kemball, *J. Catal.*, **1**, 255 (1962).
9. J.J. Rooney, *J. Catal.*, **2**, 53 (1963).
10. J.J. Rooney, *Chem. Ber.*, **2**, 242 (1966).
11. C. Kemball, *Catal. Rev.*, **5**, 33 (1971).

12. C.G. MacDonald and J.S. Shannon, *Aust. J. Chem.*, **18**, 1009 (1965).

13. J.L. Garnett and R.S. Kenyon, *Aust. J. Chem.*, **27**, 1023, 1033 (1974), and references.

14. J.L. Garnett and R.S. Kenyon, *J. Chem. Soc., Chem. Commun.*, **1971**, 1227.

15. P.A. Kramer and C. Masters, *J. Chem. Soc., Dalton Trans.*, **1975**, 849.

16. R.J. Hodges, D.E. Webster, and P.B. Wells, *J. Chem. Soc. A*, **1971**, 3230.

17. R.J. Hodges, D.E. Webster, and P.B. Wells, *J. Chem. Soc., Dalton Trans.*, **1972** , 2571, 2577.

18. S.L. Regen, *J. Org. Chem.*, **39**, 260 (1974).

19. R.R. Fraser and R.N. Renaud, *J. Am. Chem. Soc.*, **88**, 4365 (1966).

20. V. Kubelka and M. Kraus, *Collect. Czech. Chem. Commun.*, **34**, 2895 (1969).

21. J.L. Garnett, *Catal. Rev.*, **5**, 229 (1971); R.B. Moyes and P.B. Wells, *Adv. Catal.*, **23**, 121 (1973).

22. A. Farkas and L. Farkas, *Proc. R. Soc. London, Ser. A.*, **144**, 467, 481 (1934); J. Horiuti and M. Polanyi, *Nature*, **132**, 819, 931 (1933).

23. G. Fischer and M. Puza, *Synthesis*, **1973**, 218.

24. J.L. Garnett and R.J. Hodges, *Chem. Commun.*, **1967**, 1001.

25. K.P. Davis, J.L. Garnett, and J.H. O'Keefe, *Chem. Commun.*, **1970**, 1672.

26. E.K. Barefield, G.W. Parshall, and F.N. Tebbe, *J. Am. Chem. Soc.*, **92**, 5234 (1970).

27. J. Chatt and R.S. Coffey, *J. Chem. Soc. A*, **1969**, 1963.

28. J.L. Garnett, M.A. Long, R.F.W. Vining, and T. Mole, *J. Chem. Soc., Chem. Commun.*, **1972**, 1172.

3. Carbon-carbon hydrogenolysis

In general, hydrogenolytic cleavage of carbon-carbon σ-bonds requires drastic reaction conditions[1-3] and is therefore of little value as a laboratory procedure. Exceptions are the hydrogenolyses of allyl- and benzyl-substituted cyclohexadienones, e.g.[4]

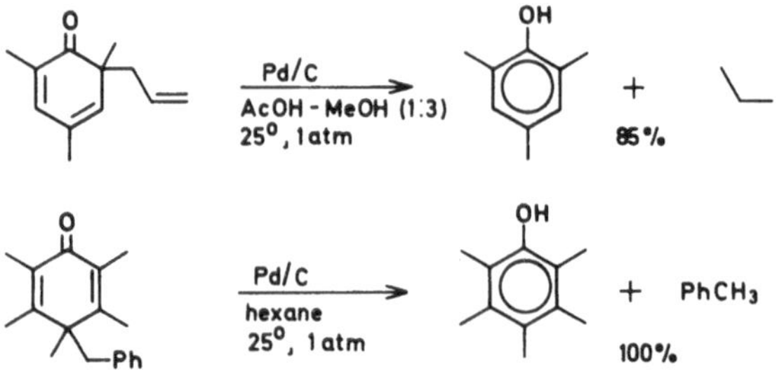

Both the presence of the π-bonds adjacent to the C-C bond to be cleaved and the concomitant aromatization of the cyclohexadienone system are responsible for this facile C-C hydrogenolysis.

Activated and/or strained cyclobutane derivatives[3,5] may also be smoothly hydrogenolyzed, e.g.

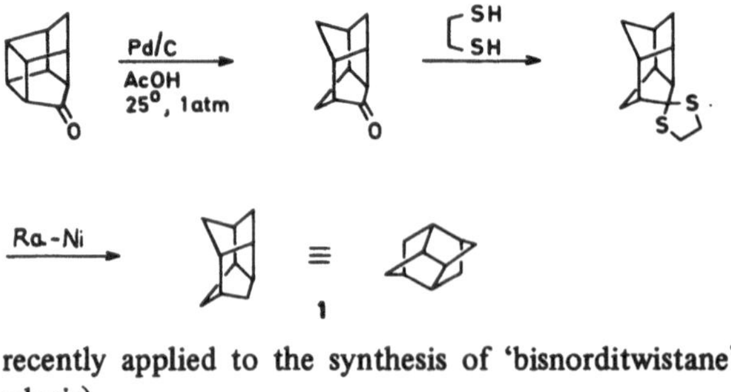

as recently applied to the synthesis of 'bisnorditwistane'[6] (1, cf. C-S hydrogenolysis).

Another (and synthetically much more valuable) exception is the hydro-

102

genolysis of cyclopropane derivatives. The ring strain together with the enhanced *p*-character of the carbon-carbon σ-bonds of the three-membered ring permits reductive cleavage to occur under mild conditions[3,7,8] Thus, the energies of activation for the hydrogenolysis of the carbon-carbon bond of cyclopropanes are 20–40 kcal/mole lower than those of other carbon-carbon bonds[9].

CATALYSTS

The order of activity of the various metals towards the hydrogenolysis of cyclopropane is Rh>Pt>Pd>Ir>Os>Ru[9].

However, in the case of conjugated cyclopropane derivatives, palladium most often shows the highest activity. Furthermore, palladium will be preferred if hydrogenation of aryl and carbonyl groups has to be avoided. Rhodium and platinum often yield mixtures in these instances and are applied preferably for the hydrogenolysis of saturated cyclopropane derivatives. Ruthenium has a

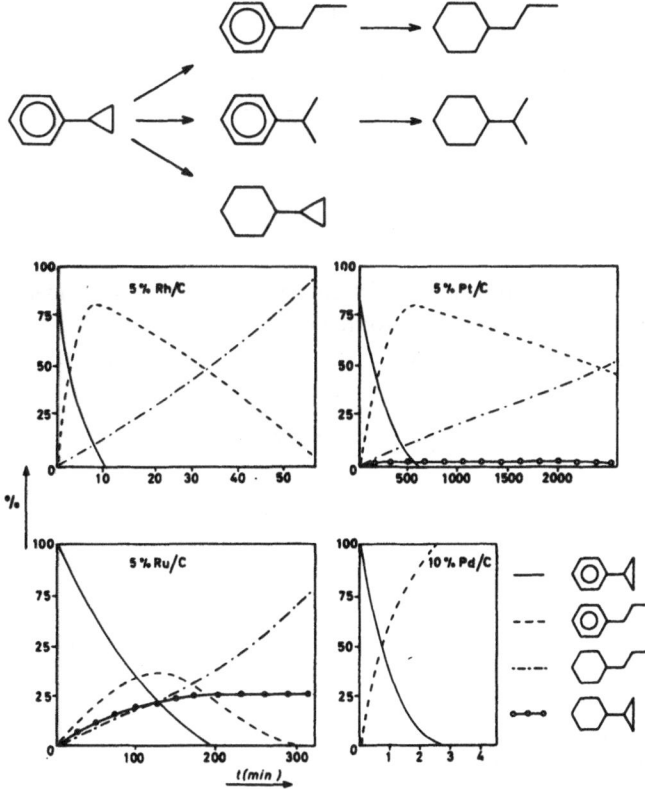

Hydrogenolysis of phenylcyclopropane (80 mg) in EtOH (20 ml) at 30° and 1 atm with 40 mg of catalyst.

103

rather low activity for the hydrogenolytic cleavage of carbon-carbon bonds.

The characteristics of the above-mentioned catalysts are nicely demonstrated in the hydrogenation/hydrogenolysis phenomena of phenylcyclopropane[10]. Whereas rhodium and platinum gave approximately 80% of propylbenzene, palladium showed 100% selectivity towards hydrogenolysis of the cyclopropane ring. With ruthenium as the catalyst, saturation of the aromatic ring occurred in competition with cyclopropane hydrogenolysis (25%). Similar phenomena have been observed for the hydrogenation/hydrogenolysis of cyclopropylalkenes[11]. Here, hydrogenolysis of the cyclopropane ring can be avoided by the use of chlorotris(triphenylphosphine)rhodium(I) as the catalyst, giving 70–97% of the corresponding cyclopropylalkanes[12]. Clearly, preferential coordination of the catalyst takes place at the carbon-carbon double bond $(K_1 \gg K_2)$:

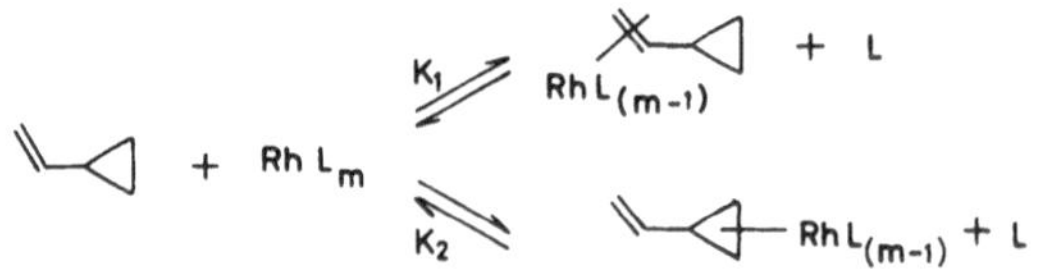

whilst no neighbouring metal atoms are present to catalyze simultaneously the carbon-carbon hydrogenolysis reaction as would occur using rhodium metal as the catalyst.

MECHANISM

It is assumed that cyclopropane adsorption occurs to give a σ-bonded precursor, that ring opening involves the formation of both 1-monoadsorbed and 1,3-diadsorbed species and that these species are converted to the alkane by addition of hydrogen[13,14].

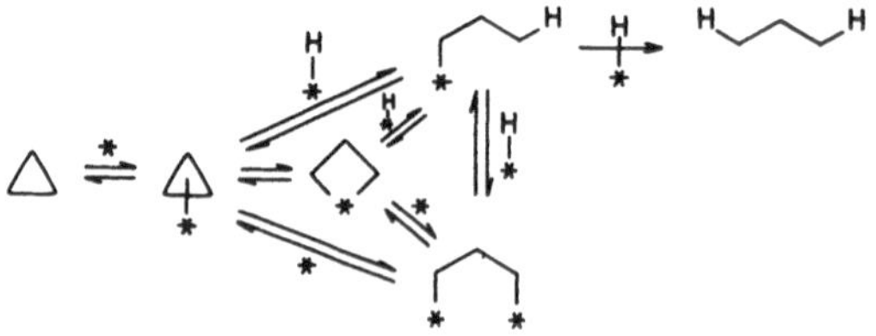

From deuterolysis experiments it appears that the formation of π-allyl intermediates (2) does not take place[13,14].

2

As a rule, the mechanistic concepts for the carbon-carbon double bond hydrogenation may be applied to the description of the carbon-carbon hydrogenolytic cleavage. In doing this, the carbon-carbon σ-bond of the double bond has to be substituted by a methylene bridge ('homo-alkenes').

STEREOCHEMISTRY

The stereochemistry of cyclopropane ring hydrogenolysis depends on the occurrence of the following possible modes of hydrogen attack on the two carbon atoms[15]:
suprafacial hydrogen attack, in which the configuration is retained (retention) or inverted at both carbon atoms:

antarafacial hydrogen attack, in which the configuration is retained at one carbon atom and inverted at the other:

Preferential suprafacial (ret, ret) deuterium attack is found for the deuterolysis of 3-acetylnortricyclene (3) over platinum[16].

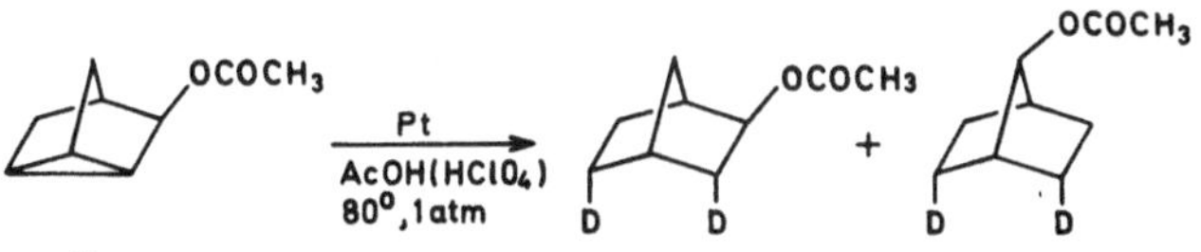

Hydrogenolysis of the less strained trans-1,2-dimethyl-1,2-diphenylcyclopropane (4) over platinum, rhodium and nickel, however, occurs by both suprafacial and antarafacial hydrogen addition[15].

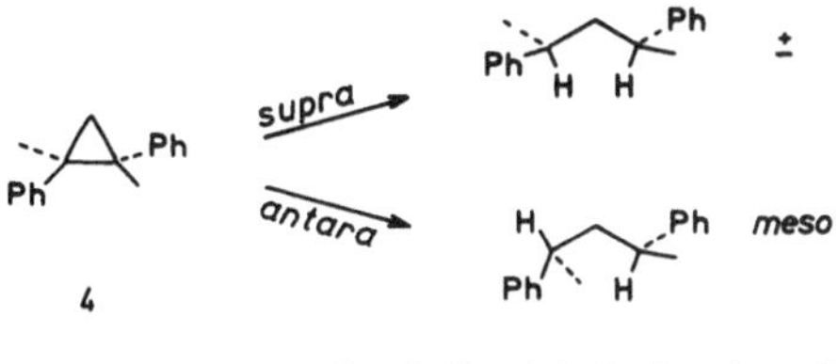

Palladium gave 80% of the (±)-derivative, i.e. preferential suprafacial hydrogen addition. With this catalyst, however, up to 80% of the meso-compound was formed if a strong acid was added to the solvent. The change in stereo-

105

chemical behaviour of the reaction (from supra- towards antarafacial) is due to a solvent participation[15] like that described earlier for the hydrogenation of the carbon-carbon double bond over palladium[17] (p 39). In acidic medium, a proton from the solvent and a hydride from the palladium both attack the carbon-carbon bond, while at the same time a neighbouring hydrogen atom on the catalyst surface goes into solution as a proton:

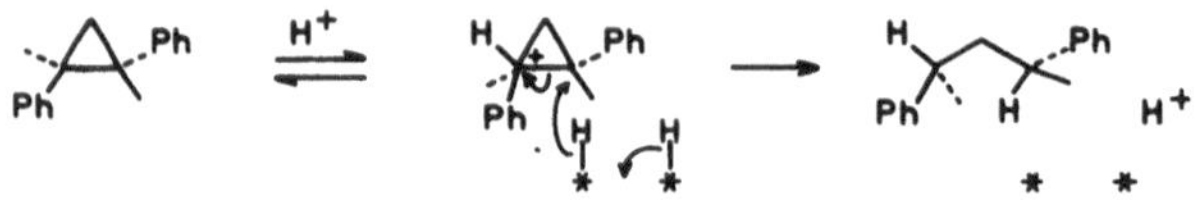

The reaction sequence closely resembles that depicted for the hydrogenolytic cleavage of the benzyl oxygen bond over palladium[18] (p 120).

In most instances, the mode of hydrogen addition to the carbon atoms is not reflected in the products formed because of the lack of sufficient substituents at *both* carbon atoms of the σ-bond to be cleaved. Stereospecificity is then described in terms of *inversion/retention* at one of the carbon atoms by analogy with the hydrogenolysis of carbon-hetero bonds.

Several examples are given below:

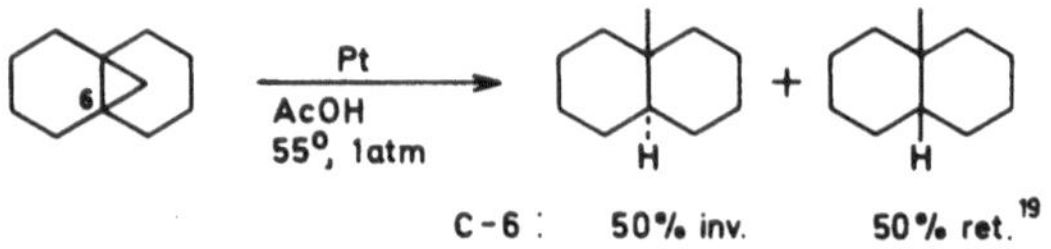

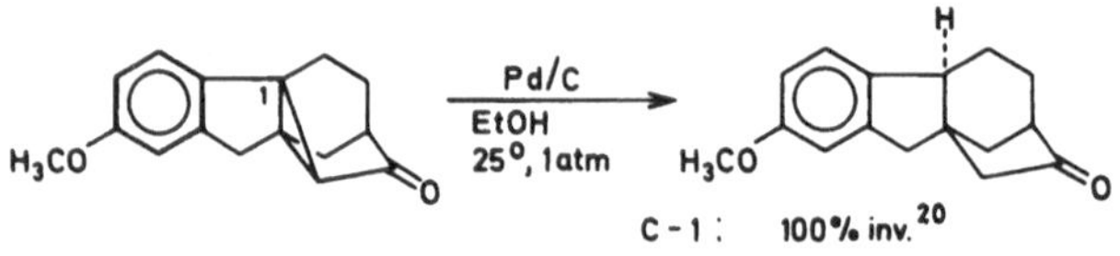

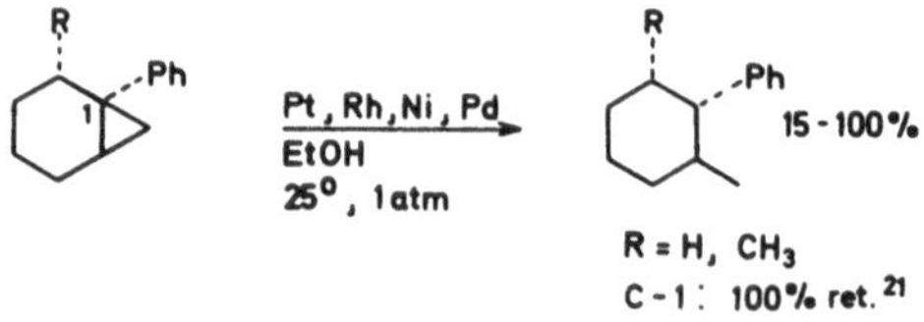

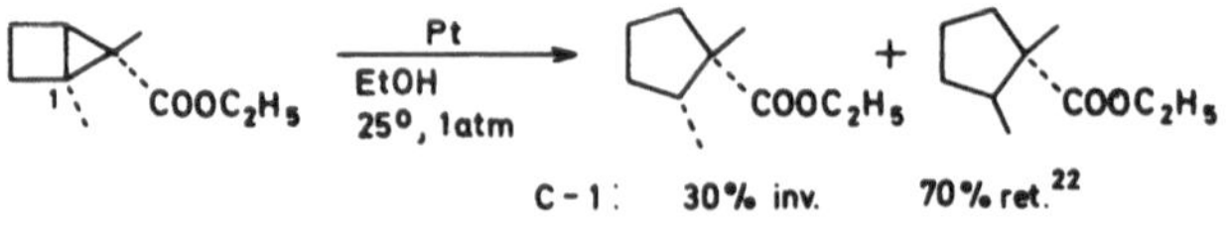

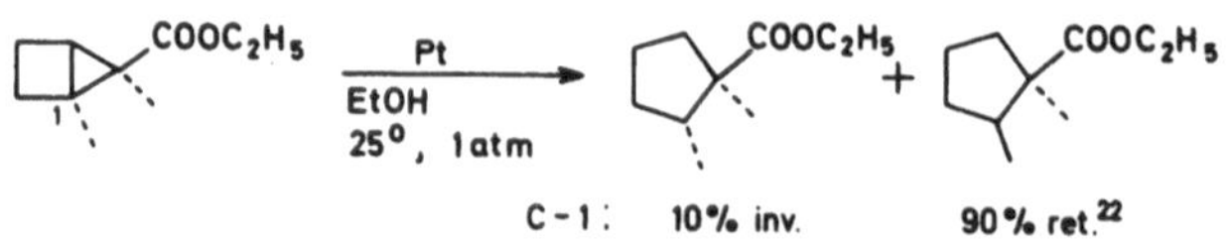

106

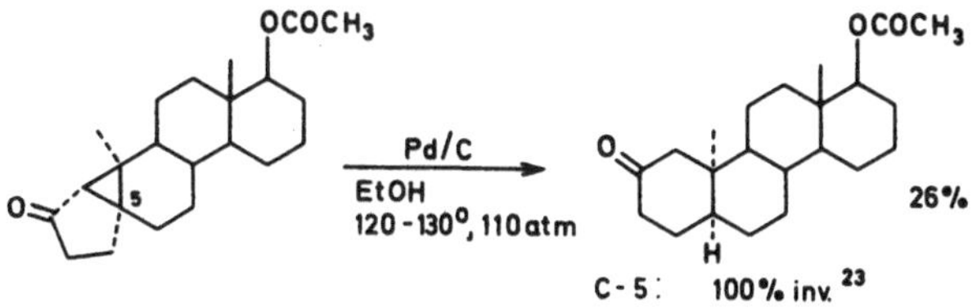

Pd/C
EtOH
120-130°, 110 atm

26%

C-5 : 100% inv.[23]

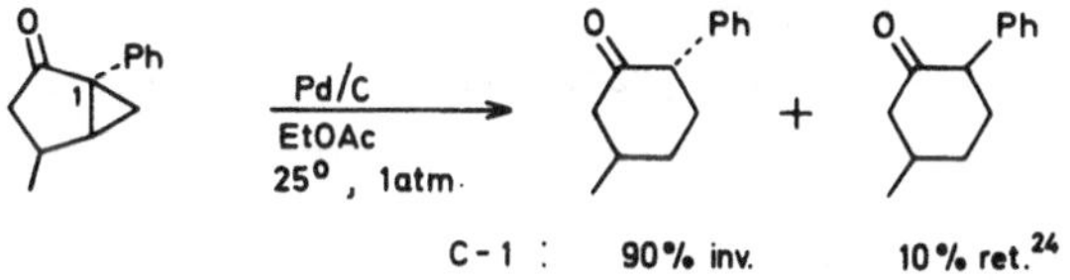

Pd/C
EtOAc
25°, 1 atm

C-1 : 90% inv.   10% ret.[24]

More data is required in order to give a general rule concerning the stereo-chemistry of the cyclopropane ring hydrogenolysis. Furthermore, the variation in stereochemical course over palladium may be due to the presence of traces of acids as cited above[15].

REGIOSELECTIVITY

The situation with respect to the regioselectivity of the reaction is clearer. First, hydrogenolysis of alkylcyclopropanes involves the preferential cleavage of the least-substituted or -hindered carbon-carbon bond.

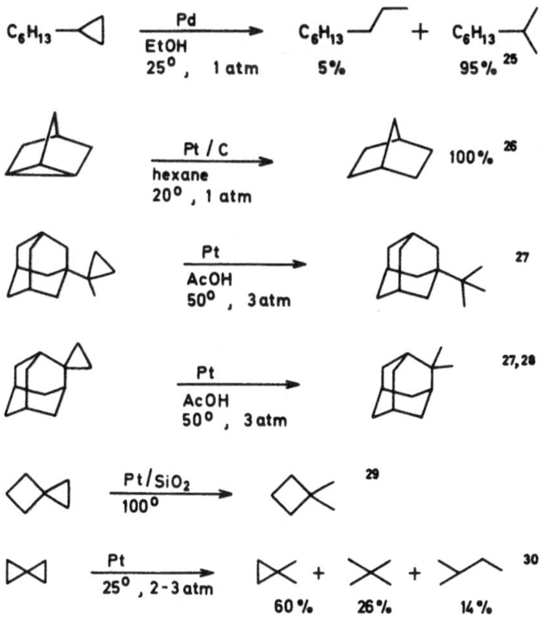

C_6H_{13} ◁
Pd
EtOH
25°, 1 atm

C_6H_{13} ───/   + C_6H_{13} ───<
5%   95%[25]

Pt / C
hexane
20°, 1 atm
100%[26]

Pt
AcOH
50°, 3 atm
[27]

Pt
AcOH
50°, 3 atm
[27,28]

Pt/SiO₂
100°
[29]

Pt
25°, 2-3 atm
+ +   [30]
60%   26%   14%

107

The hydrogenolysis of benzylcyclopropane (5) is an exception[25].

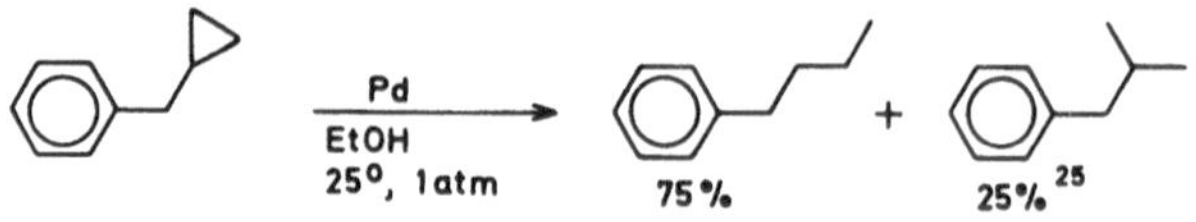

75%   25%[25]

5

This can be understood by considering the two possible ways of chemisorption of both the benzene and the cyclopropane ring on the catalyst surface

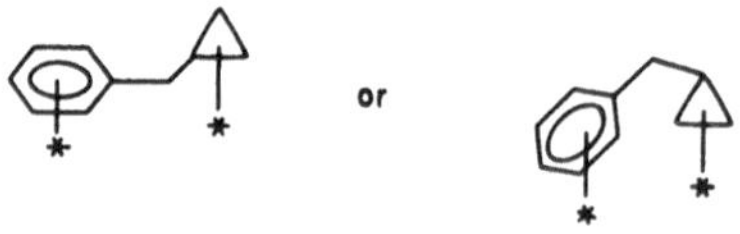

or

leading to the cleavage of the C(1)-C(2) and C(2)-C(3) bonds, respectively. The left-hand structure will be favoured because of the optimal flat adsorption of the system without any internal van der Waals repulsion.

Second, vinyl-, acyl-, oxycarbonyl- and aryl-substituted cyclopropanes show preferred cleavage of the adjacent carbon-carbon bond, as may be seen from the following examples with *palladium* as the catalyst and from the examples given earlier (p 106).

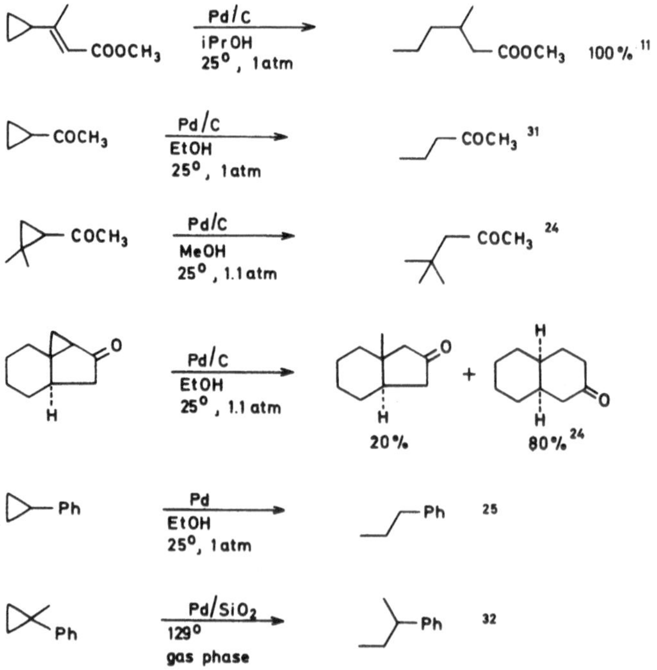

108

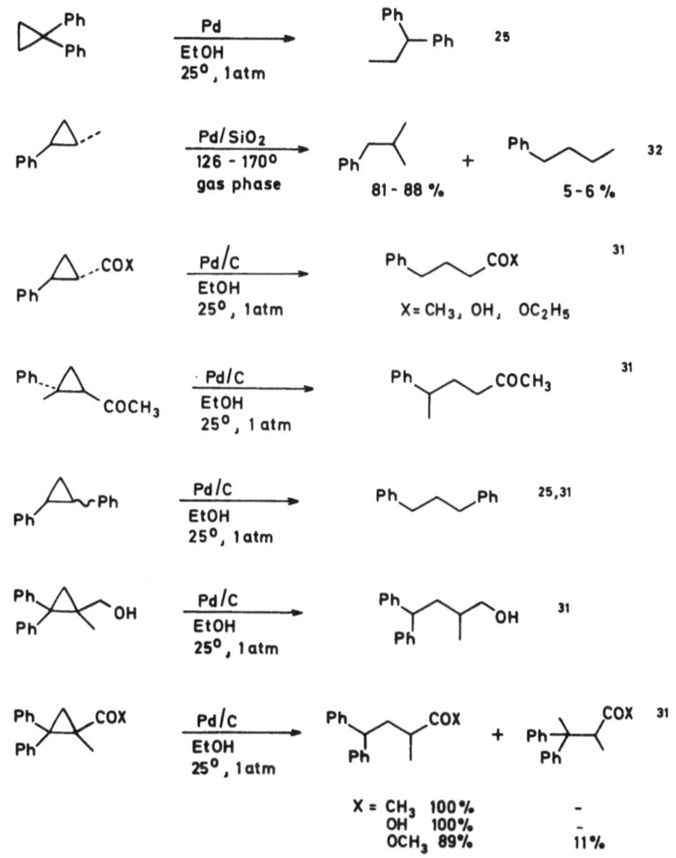

Ph — Pd / EtOH / 25°, 1atm → Ph / Ph 25

Ph △ — Pd/SiO₂ / 126 - 170° / gas phase → Ph (81 - 88 %) + Ph (5-6 %) 32

Ph △ ..COX — Pd/C / EtOH / 25°, 1atm → Ph COX 31
X = CH₃, OH, OC₂H₅

Ph.. △ COCH₃ — Pd/C / EtOH / 25°, 1 atm → Ph COCH₃ 31

Ph △ Ph — Pd/C / EtOH / 25°, 1atm → Ph Ph 25,31

Ph Ph △ OH — Pd/C / EtOH / 25°, 1atm → Ph Ph OH 31

Ph Ph △ COX — Pd/C / EtOH / 25°, 1atm → Ph Ph COX + Ph Ph COX 31
X = CH₃ 100 % —
OH 100 % —
OCH₃ 89 % 11 %

The unsaturated 'handle' at the cyclopropane ring will force the chemisorption towards:

X—⫶

and will thus promote the hydrogenolysis of the adjacent carbon-carbon bond. In most instances, the above-mentioned functional group effect overrules the opposite effect caused by alkyl substituents.

REACTIVITY

The rate of hydrogenolysis is enhanced by the addition of a strong acid, probably by protonation of the cyclopropane ring. Both electronic and steric effects of the substituents influence the reactivity considerably. The following sequences over palladium are known[24,31]:

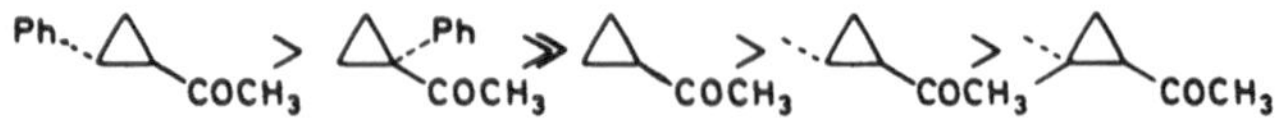

Ph—△(COCH₃) > △(Ph)(COCH₃) ≫ △(COCH₃) > △(COCH₃) > △(COCH₃)

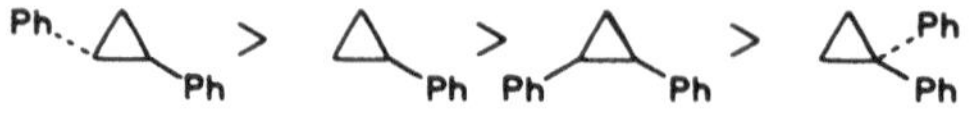

Ph—△(Ph) > △(Ph) Ph—△(Ph) > △(Ph) > △(Ph)(Ph)

The decrease in rate of hydrogenolysis parallels the decrease in conjugation between the cyclopropane ring and the substituent[32]. Alkyl substituents retard the reaction by steric interaction between the alkyl substituents and the catalyst surface. Under conditions where alkylcyclopropanes are stable, the conjugated cyclopropanes show rapid ring cleavage. A clear picture concerning the electronic nature of both the initial adsorbed state and the transition state of the cyclopropane ring hydrogenolysis has not yet been established[25,31-33]. Recently, the occurrence of metallocarbonium ions in the hydrogenolysis of nortricyclenes over palladium has been assumed, by analogy with the homogeneously catalyzed isomerization of cyclopropane derivatives, in order to explain some observed hydrogen and methyl shifts[34,35].

SELECTIVITY

In particular with palladium as the catalyst, high selectivity towards carbon-carbon hydrogenolysis is obtained in the presence of other reducible groups such as aryl and ketone. Some synthetic examples involving cyclopropane ring hydrogenolysis are depicted below.

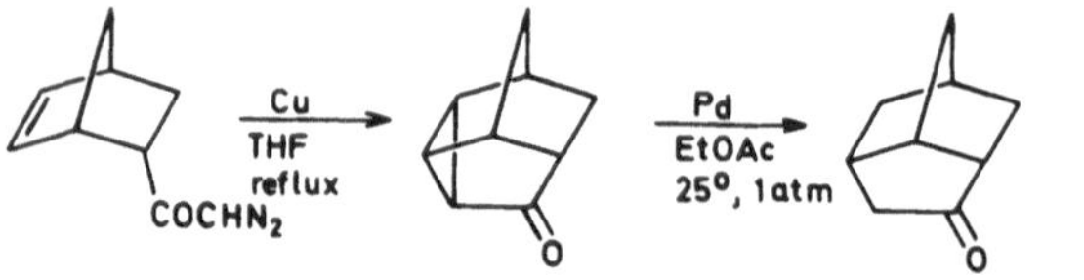

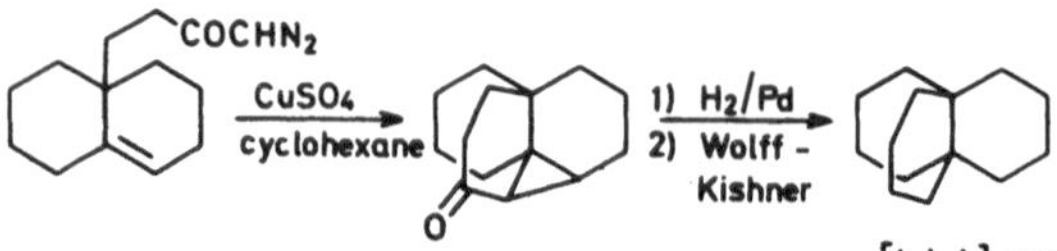

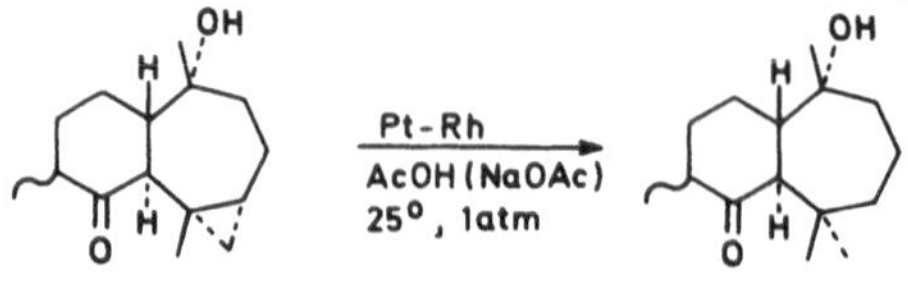

Cyclopropanation[39] of the alkenic precursors and subsequent hydrogenolysis[27,28] may be extremely useful for the synthesis of compounds containing isopropyl, *gem*-dimethyl, *t*-butyl, or angular methyl groups.

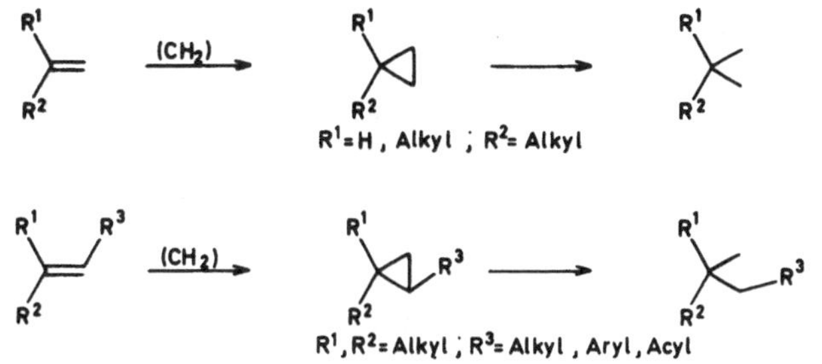

$R^1 = H$, Alkyl ; $R^2 =$ Alkyl

$R^1, R^2 =$ Alkyl ; $R^3 =$ Alkyl, Aryl, Acyl

Cyclopropenes may be cleaved by simultaneous hydrogenation/hydrogenolysis of the three-membered ring[40,41].

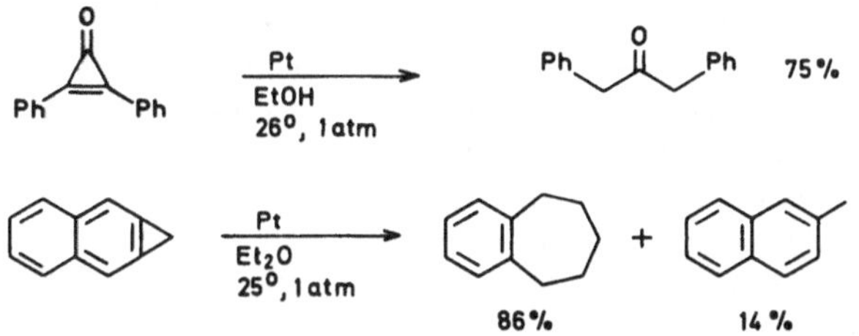

Bicyclobutanes are directly hydrogenolyzed to the corresponding tetrahydro derivatives[42].

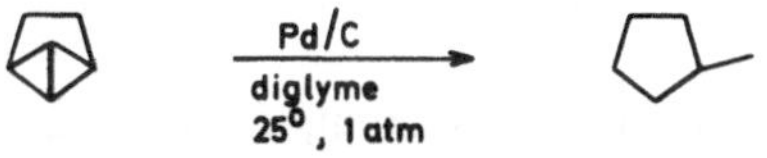

Cleavage of the two bonds takes place during just one residence of the molecule on the catalyst[42,43]. In one particular case (6), an unusual dihydro derivative was selectively formed[44].

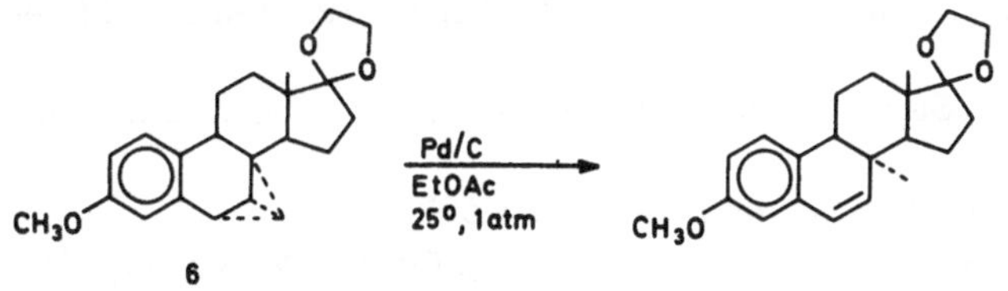

This indicates that dihydro intermediates (π- or 1,2-di-σ-bonded on the catalyst) may play a role in the hydrogenolysis of simple bicyclobutanes. Probably the low rates of desorption of these intermediates (*i.e.* strong chemisorption) prevent their detection in solution.

References

1. J.H. Sinfelt, *Catal. Rev.*, 3, 175 (1969).

2. J.H. Sinfelt, *Adv. Catal.*, **23**, 91 (1973).

3. J. Newham, *Chem. Rev.*, **63**, 123 (1963).

4. B. Miller and L. Lewis, *J. Org. Chem.*, **39**, 2605 (1974), and references.

5. *Cf.* N.A. Sasaki, R. Zunker, and H. Musso, *Chem. Ber.*, **106**, 2992 (1973); F.R. Jensen and W.E. Coleman, *J. Am. Chem. Soc.*, **80**, 6149 (1958); E. Osawa, P. von R. Schleyer, L.W.K. Chang, and V.V. Kane, *Tetrahedron Lett.*, **1974**, 4189; H. Musso, *Chem. Ber.*, **108**, 337 (1975).

6. K. Hirao, T. Iwakuma, M. Taniguchi, E. Abe, and O. Yonemitsu, *J. Chem. Soc., Chem. Commun.*, **1974**, 691.

7. M. Yu Lukina, *Russ. Chem. Rev.* (Engl. Transl.), **31**, 419 (1962).

8. D. Wendisch, in *Methoden der Organische Chemie (Houben-Weyl)*, 4/3, G. Thieme Verlag, Stuttgart, 1971, p 656.

9. R.A. Dalla Betta, J.A. Cusumano, and J.H. Sinfelt, *J. Catal.*, **19**, 343 (1970).

10. J.S. Winnik, A.P.G. Kieboom, and H. van Bekkum, unpublished results.

11. S.R. Poulter and C.H. Heathcock, *Tetrahedron Lett.*, **1968**, 5339, 5343.

12. C.H. Heathcock and S.R. Poulter, *Tetrahedron Lett.*, **1969**, 2755.

13. J.R. Anderson and N.R. Avery, *J. Catal.*, **8**, 48 (1967).

14. J.A. Roth, *J. Am. Chem. Soc.*, **92**, 6658 (1970).

15. A.P.G. Kieboom, A.J. Breijer, and H. van Bekkum, *Rec. Trav. Chim. Pays-Bas*, **93**, 186 (1974).

16. M.N. Akhtar and W.R. Jackson, *J. Chem. Soc., Chem. Commun.*, **1972**, 813.

17. F. van Rantwijk, A. van Vliet, and H. van Bekkum, *J. Chem. Soc., Chem. Commun.*, **1973**, 234.

18. A.P.G. Kieboom, J.F. de Kreuk, and H. van Bekkum, *J. Catal.*, **20**, 58 (1971).

19. Z. Majerski and P. von R. Schleyer, *Tetrahedron Lett.*, **1968**, 6195.

20. U.R. Ghatak, P.C. Chakraborti, B.C. Ranu, and B. Sanyal-Moitra, *J. Chem. Soc., Chem. Commun.*, **1973**, 548; P.N. Chakrabortty, R. Dasgupta, S.K. Dasgupta, S.R. Ghosh, and U.R. Ghatak, *Tetrahedron*, **28**, 4653 (1972).

21. S. Mitsui, Y. Sugi, M. Fujimoto, and K. Yokoö, *Tetrahedron*, **30**, 31 (1974).

22. M.J. Jorgenson, *Tetrahedron Lett.*, **1968**, 4577.

23. R.L. Augustine and E.J. Reardon, Jr., *J. Org. Chem.*, **39**, 1627 (1974).

24. R. Fraisse-Jullien, C. Frejaville, V. Toure, and M. Derieux, *Bull. Soc. Chim. Fr.*, **1968**, 4444.

25. W.J. Irwin and F.J. McQuillin, *Tetrahedron Lett.*, **1968**, 2195.

26. I.I. Khochenko, E.M. Mil'vitskaya, and A.F. Platé, *J. Org. Chem. USSR* (Engl. Transl.), **8**, 560 (1972).

27. C.W. Woodworth, V. Buss, and P. von R. Schleyer, *Chem. Commun.*, **1968**, 569.

28. V. Buss, R. Gleiter, and P. von R. Schleyer, *J. Am. Chem. Soc.*, **93**, 3927 (1971).

29. N.S. Nametkin, V.M. Vdovin, E.S. Finkelshtein, A.M. Popov, A.V. Egorov, *Izv. Akad. Nauk SSSR, Ser. Khim*, **1973**, 2806, *Chem. Abstr.*, **80**, 82203g (1974).

30. V.A. Slabey, *J. Am. Chem. Soc.*, **69**, 475 (1947).

31. A.L. Schultz, *J. Org. Chem.*, **36**, 383 (1971).

32. J.A. Roth, *J. Catal.*, **26**, 97 (1972).

33. J. Smejkal and J. Farkas, *Collect. Czech. Chem. Commun.*, **28**, 1557 (1963).

34. M.N. Akhtar, W.R. Jackson, J.J. Rooney, and N.G. Samman, *J. Chem. Soc., Chem. Commun.*, **1974**, 155.

35. M.N. Akhtar, W.R. Jackson, and J.J. Rooney, *J. Am. Chem. Soc.*, **96**, 276 (1974).

36. A. Nickon, H. Kwasnik, T. Swartz, R.O. Williams, and J.B. DiGiorgio, *J. Am. Chem. Soc.*, **87**, 1615 (1965).

37. J. Altman, D. Becker, D. Ginsburg, and H.J.E. Loewenthal, *Tetrahedron Lett.*, **1967**, 757.

38. B.D. Challand, G. Kornis, G.L. Lange, and P. de Mayo, *Chem. Commun.*, **1967**, 704.

39. H.E. Simmons, T.L. Clairns, S.A. Vladuchick, and C.M. Hoiness, *Org. React.*, **20**, 1 (1973).

40. R. Breslow, T. Eicher, A. Krebs, R.A. Peterson, and J. Posner, *J. Am. Chem. Soc.*, **87**, 1320 (1965).

41. W.E. Billups and W.Y. Chow, *J. Am. Chem. Soc.*, **95**, 4099 (1973).

42. D.M. Lemal and K.S. Shim, *Tetrahedron Lett.*, **1964**, 3231.

43. K.B. Wiberg, *Adv. Alicyclic Chem.*, **2**, 185 (1968).

44. E. Galantay, N. Paolella, S. Barcza, R.V. Coombs, and H.P. Weber, *J. Am. Chem. Soc.*, **92**, 5771 (1970).

4. Carbon-oxygen hydrogenolysis

Hydrogenolysis of the carbon-oxygen bond occurs readily when the carbon atom is directly linked to or is part of an aromatic system (as in the case of benzyl alcohol and phenol derivatives[1]) in particular with palladium as the catalyst. Further, oxiranes are cleaved smoothly whereas hydrogenolysis of acetals and activated esters has been observed in some instances. Preparative applications in organic synthesis consist in both the selective removal of oxygen and nitrogen protective groups (*e.g.* in peptide synthesis), in the conversion of aryl carbonyl into aryl methylene groups, in the reductive alkylation of alcohols and amines, in the synthesis of chiral hydrocarbons from the corresponding oxygen derivatives, and in the replacement of phenolic hydroxyl groups by hydrogen. The hydrogenolysis reaction of sp^3- and sp^2-hybridized carbon-oxygen bonds will be dealt with consecutively below.

(sp^3)CARBON-OXYGEN BONDS

Alcohols and ethers

Aliphatic carbon-oxygen compounds such as alcohols and ethers require extreme reaction conditions. However, partial hydrogenolytic cleavage of the carbon-oxygen bond at ambient temperature has been observed for alkyl trimethylsilyl ethers (1)[2].

$$R-O-SiMe_3 \xrightarrow[\substack{cyclohexane \\ 25^\circ, 1\,atm.}]{Pd/C} R-H \ + \ HOSiMe_3$$
$$1$$

In addition, *O*-alkylisoureas (2) can be hydrogenolyzed to the respective hydrocarbon and urea under mild reaction conditions[3].

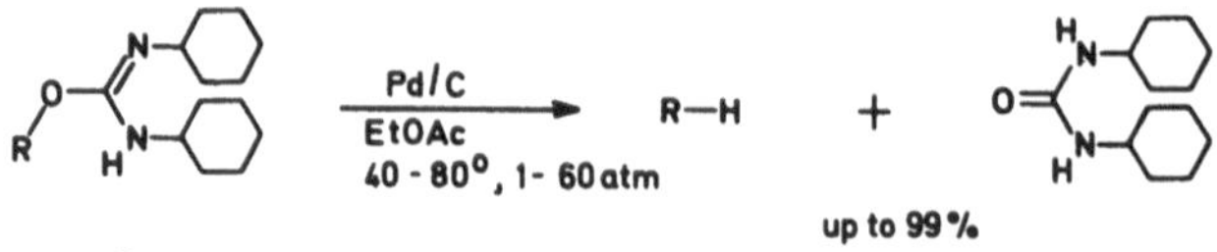

114

Since these ethers are conveniently obtained from the alcohol and *N,N*-dicyclo-hexylcarbodiimide, the reaction is useful for the reductive removal of aliphatic hydroxyl groups. The enhanced reactivity of the *O*-alkylisoureas will be due to both the stronger chemisorption (resulting in a closer contact between the C-O bond and the catalyst) and the better leaving nature of the hetero-ether system (by the formation of the urea product 3) as compared with a hydroxyl or an alkoxyl group in aliphatic alcohols and ethers, respectively.

A similarly enhanced reactivity has been observed for the hydrogenolysis of aryl- and carbonyl-oxygen bonds when the hydroxylic compounds are con-verted to hetero(cyclic) ether or ester derivatives respectively (pp 125–128).

Homobenzyl oxygen derivatives (4), in addition to the easily hydrogeno-lyzable benzylalcohol compounds (see p 118), also show reductive carbon-oxygen cleavage over Raney nickel or palladium catalysts[4].

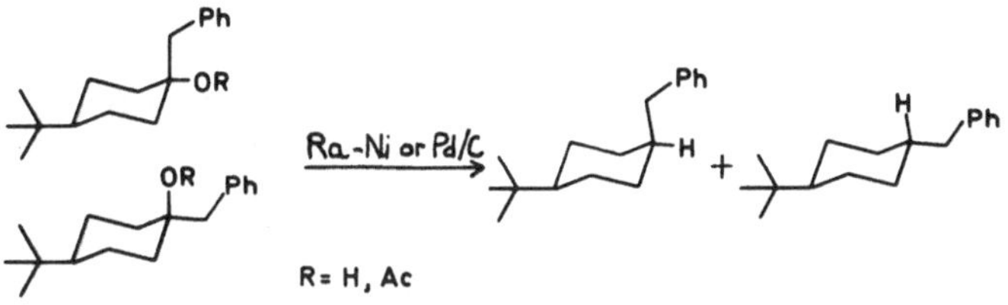

The reaction proceeds through the desorbed alkene

arising from the homobenzyl oxygen compounds by elimination of OR and a benzylic hydrogen:

With palladium as the catalyst, the intermediate exocyclic alkene mainly isomerizes to the more stable compound

before saturation, as would be expected from carbon-carbon double bond hydrogenation phenomena on this catalyst (see p 41).

Acetals

Acetals (5) are found to be converted, at somewhat higher temperature, to the corresponding ethers, with rhodium as the catalyst in acidic medium[5].

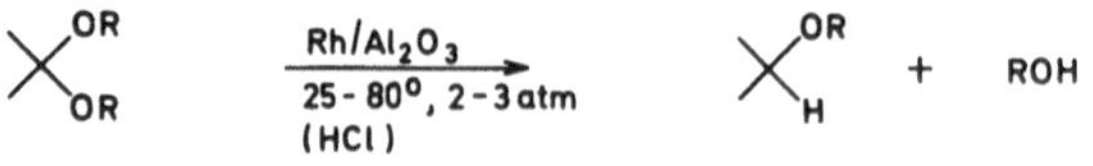

R = CH$_3$, C$_4$H$_9$, i-C$_3$H$_7$, c-C$_6$H$_{11}$

5

Hydrogenation of aliphatic ketones over platinum in acid alcoholic medium affords ethers in high yield[6], probably *via* hydrogenolytic cleavage of the carbon-hydroxyl bond of the intermediate hemiacetal **6**.

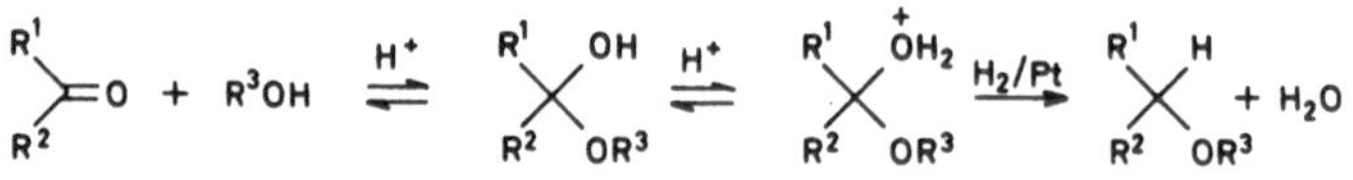

6

The presence of acid is essential; both the formation of the hemiacetal and its hydrogenolysis reaction are favoured by proton catalysis. The latter reaction may be described in terms of a nucleophilic displacement (*cf.* the benzyl-oxygen hydrogenolysis, p 120) according to

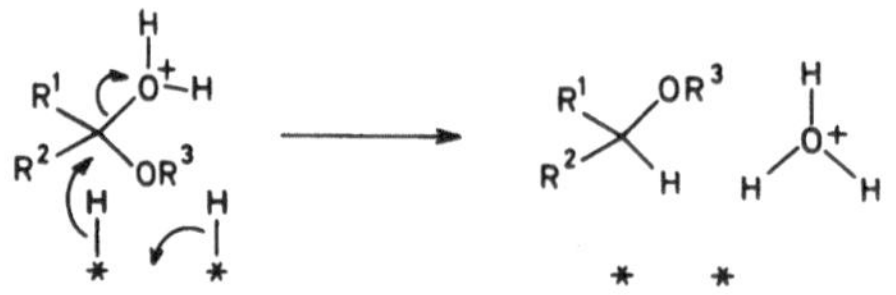

(A similar displacement would seem to occur for the above-mentioned hydrogenolysis of acetals over rhodium). Protonation of the hemiacetal results in the formation of a better leaving group ($-O^+H_2$ with respect to $-OH$). Some results[6] of this reaction are summarized in table VI.

Table VI   Yield of ether (%)[a]

Ketone	Alcohol				
	MeOH	EtOH	PrOH	iPrOH	c-Hexanol
isopropyl methyl ketone	93	75	90-92	58	
cyclopentanone	84	79-90	92	80	
cyclohexanone	95	61-67	85	52	39

a: Pt, 25°, 1 atm, with the alcohol as the solvent (in the presence of HCl).

Reductive alkylation of amines

By analogy with the above-mentioned ether formation, hydrogenation of aldehydes and ketones in the presence of amines affords *secondary* and *tertiary* amines.

116

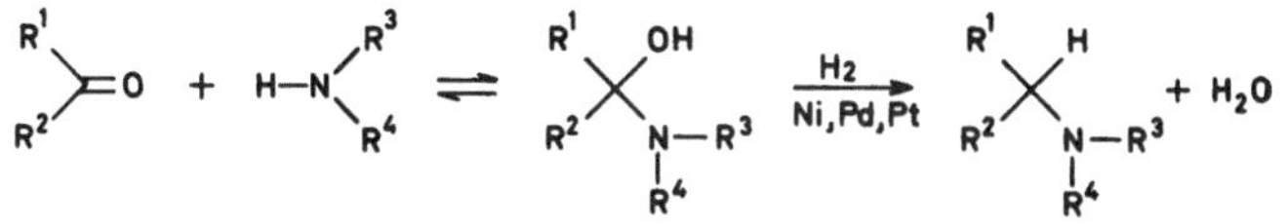

The reaction is commonly denoted as the reductive alkylation of amines[7] and is a convenient alternative to the Leuckart reaction[8], in particular for aliphatic aldehydes and ketones with lower molecular weight for which the latter reaction results in substantial amounts of undesired condensation products. The reaction between aliphatic ketones and *primary* alkylamines over palladium at approximately 100–200° and 50–140 atm gave the corresponding dialkylamines in 50–98% yield, usually 90%[9]. The use of formaldehyde offers a convenient method for the direct methylation of both aliphatic[10] and aromatic[11] amines.

$$2\,H_2CO + CH_3NHCO(CH_2)_4\,NH_2 \xrightarrow[\substack{H_2O \\ 25°,\,1\,atm \quad 60\%}]{Pd} CH_3NHCO(CH_2)_4\,N(CH_3)_2 + 2\,H_2O$$

$$2\,H_2CO + 4\text{-}CH_3COC_6H_4NH_2 \xrightarrow[\substack{EtOH\,(HCl) \\ 25°,\,2\text{-}3\,atm \quad 70\%}]{Pt} 4\text{-}CH_3COC_6H_4N(CH_3)_2 + 2\,H_2O$$

Conventional methylation methods resulted in very low yields in the latter case. Reductive alkylation of aryl amines (or their nitro precursors) with aliphatic ketones over platinum metal sulphides affords the desired N-alkylarylamines[12-14].

Oxiranes

Oxiranes are hydrogenolyzed under very mild reaction conditions as would be expected by analogy with the enhanced reactivity of cyclopropane with respect to linear alkanes towards carbon-carbon cleavage (p 102). The regioselectivity depends on the medium. In neutral or basic solution the highest substituted alcohol is formed preferentially[15]:

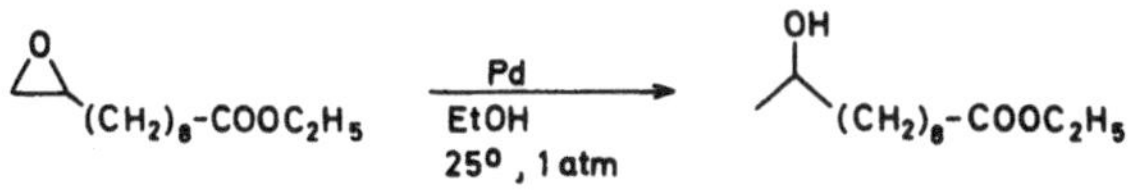

whereas an acid medium affords mainly the least substituted alcohol[16]:

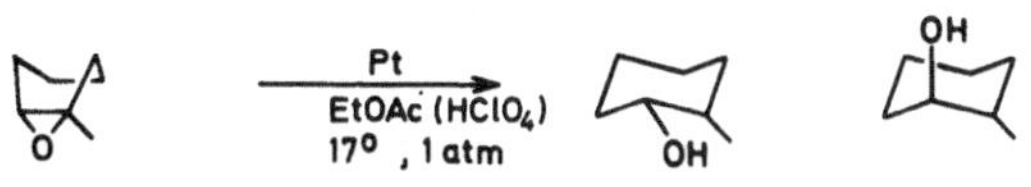

Steric effects, however, may alter the mode of ring opening.

Probably, protonation of the oxirane oxygen is involved in acid medium,

117

and this results in the occurrence of positively charged species during the hydrogenolytic cleavage process.

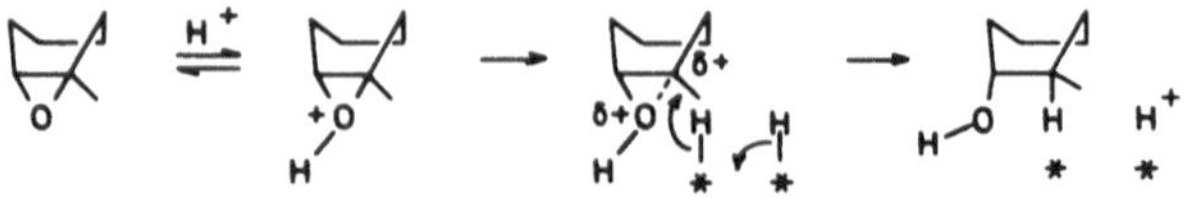

That electronic effects play an important role on the regioselectivity of the hydrogenolysis of the oxirane ring also appears from[17]:

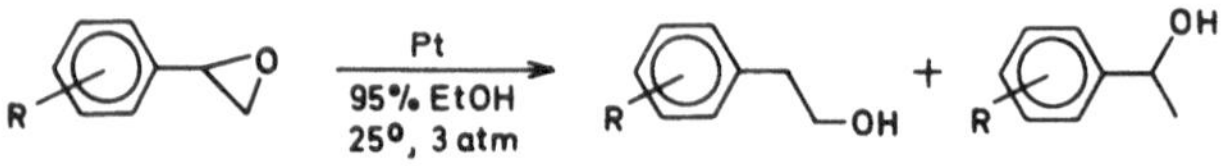

R = 4-CH$_3$	100 %	0 %
H	100 %	0 %
3- OCH$_3$	88 %	12 %
4- Br	82 %	18 % *
3,4-diCl	35 %	65 % *

(* including ethylbenzene)

The directing effect of the phenyl group towards 1,2-cleavage of the oxirane ring owing to the greater stabilization of

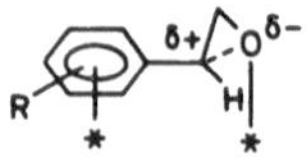

compared to

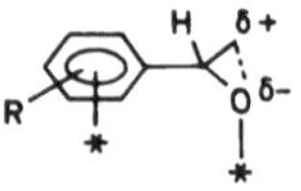

is counteracted by electron-withdrawal by the substituents.

BENZYL-OXYGEN BONDS

The last example is a particular type of the hydrogenolysis of benzyl alcohol derivatives with the general formula 7:

Numerous benzyl alcohol derivatives have been studied and have been found to react very rapidly under mild conditions. Generally, the hydrogenolysis reaction proceeds at a progressively decreasing rate, due to coadsorption of the aromatic hydrocarbon as it is formed in increasing amounts during the reaction[18-20]. The adsorption equilibrium

118

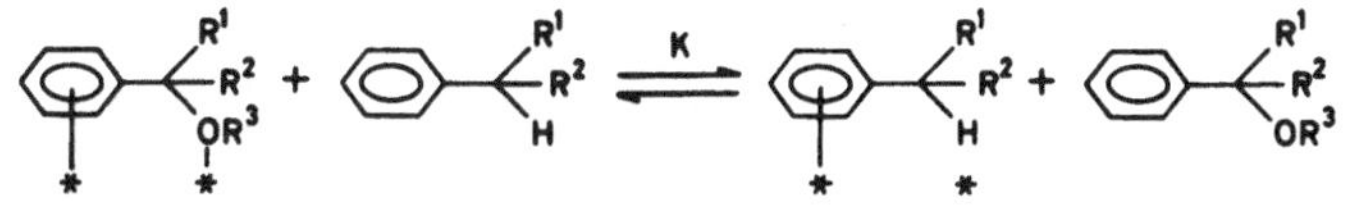

will be shifted to the right as the reaction proceeds, *i.e.* the fraction of the catalyst surface available to the benzyl alcohol derivative diminishes. Depending on the size and nature of R^1, R^2, and R^3, adsorption equilibrium constants K of 1-2 have been found on palladium[20], which implies only a small contribution of the leaving group OR^3 to the strength of adsorption. Therefore, the hydrogen uptake curve resembles closely that expected for a first order reaction. With palladium and Raney nickel catalysts, the hydrogenation of the aromatic ring is relatively slow.

The use of polar solvents (acetic acid, ethanol, methanol) facilitates the reaction[19], probably because of a better solvation of the positively charged transition state (see below). Water-miscible solvents are necessary for the conversion of benzyl alcohols in order to prevent coagulation of the catalyst by the water formed during the reaction. Palladium is by far the most active catalyst, although satisfactory results may be obtained in some cases using Raney nickel, rhodium or platinum as the catalyst. On the other hand, ruthenium should be used if hydrogenolysis of the carbon-oxygen bond is not desired, although it is difficult to keep the benzyl-oxygen bond intact[19].

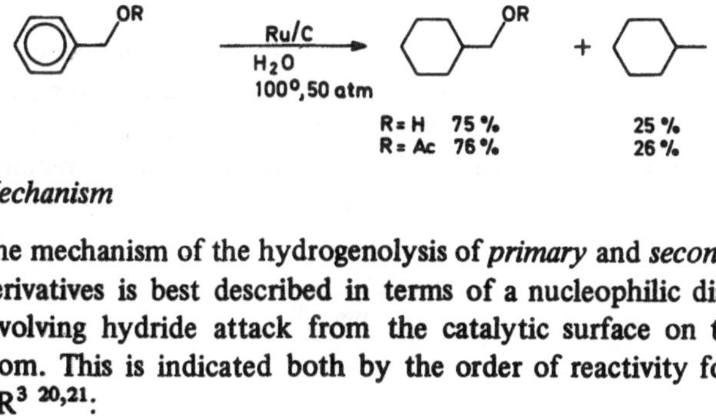

R = H 75 %	25 %
R = Ac 76 %	26 %

Mechanism

The mechanism of the hydrogenolysis of *primary* and *secondary* benzyl alcohol derivatives is best described in terms of a nucleophilic displacement reaction involving hydride attack from the catalytic surface on the benzylic carbon atom. This is indicated both by the order of reactivity for the leaving group OR^3 [20,21]:

OH, OAlkyl $\ll$ OAryl $<$ O^+H_2, $O^+HAlkyl$, OAc $<$ $OCOCF_3$, and by the similarity between the effect of R^2 on the reaction rate of hydrogenolysis and that of homogeneous S_N2 displacement[20]:

Reaction	R^2			
	Me	Et	iPr	tBu
Hydrogenolysis of R^2-Ç-OH over palladium (with Ph and H substituents)	1.0	0.47	0.031	<0.0011
S_N2 displacement of R^2-Ç-X (with H substituents)	1.0	0.4	0.03	0.000 01

119

In accordance with this picture, attempted hydrogenolysis of the carbon-oxygen bond in **8** over palladium gives exclusively 7-acetoxy[4.2.0]bicyclo-octane **(9)**[22] as would be expected from the very slow displacement reactions of cyclobutyl derivatives[23].

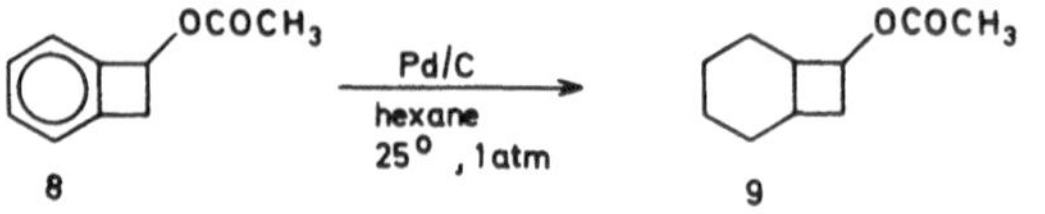

Tertiary benzyl alcohol derivatives **(10)** show a greater degree of bond-breaking than of bond-making at the transition state **(11)**, *i.e.* S_N1-type character[20]:

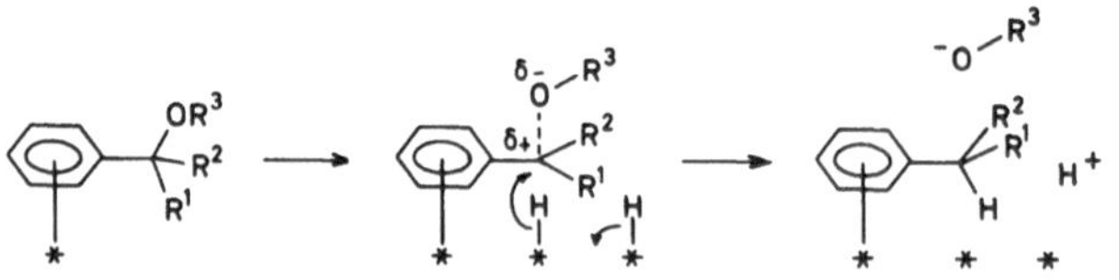

as appears from the rather negative (-1.40) Hammett ρ-value for substituted benzyl alcohols (R^1=Me, R^2=iPr) and the smaller decrease in reaction rate upon the introduction of alkyl substituents R^1 and R^2 than would be expected for an S_N2 mechanism.

It may be noted that proton catalysis is essential to the hydrogenolysis of the carbon-oxygen bond of benzyl alcohols and benzyl alkyl ethers over palla-dium[20]. Either the use of an acidic catalyst or the addition of a trace of acid is sufficient. If the reactant itself contains a basic function more acid is required in order to protonate the basic function completely[24].

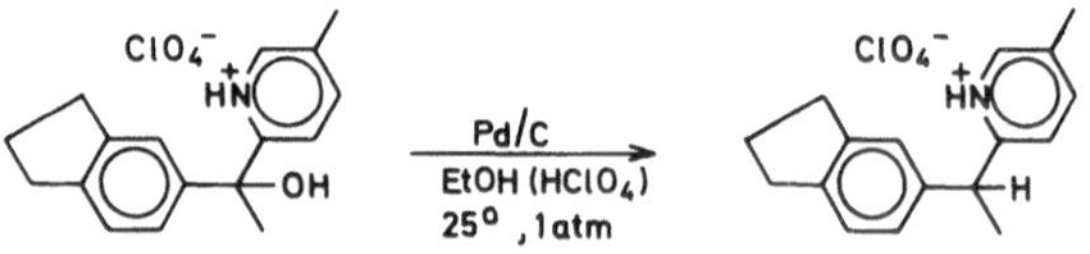

The proton catalysis may be formulated as follows[20].

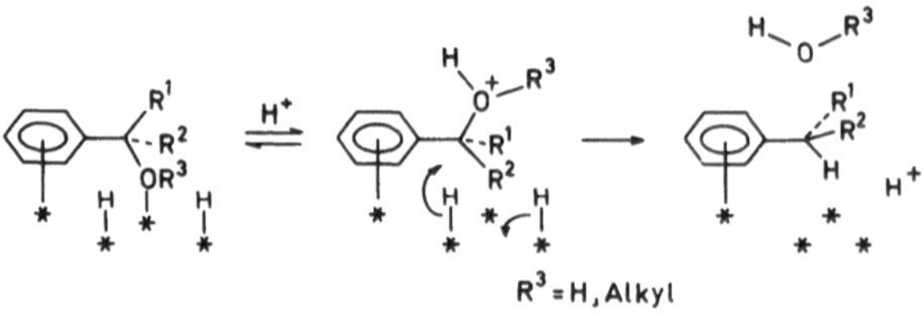

R^3 = H, Alkyl

Addition of base prevents hydrogenolysis of the carbon-oxygen bond in these compounds, showing the importance of producing a good leaving group.

120

Stereochemistry

The stereochemistry of the reaction has been investigated extensively by the use of chiral *tertiary* benzyl oxygen derivatives[25-27]. With palladium as the catalyst, *inversion* of configuration at the benzylic carbon is most often found.

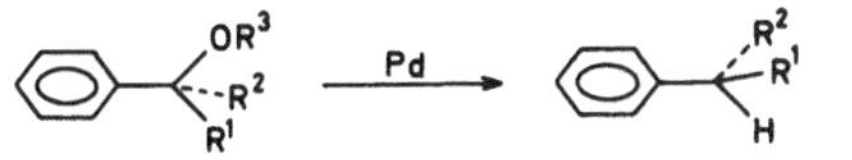

R¹	R²	R³	Inversion (%)
Me	Et	OH	91-95
Me	COOEt	OH	95-98
Me	COOEt	OMe	97-98
Me	(CH₂)ₙCOOR (n=1 or 2)	OH	84-100

The results are in line with the mode of hydride attack described above. However, with Raney nickel as the catalyst, the reaction occurs with *retention* of configuration[25-27]. Evidently, an S_Ni type mechanism is operative:

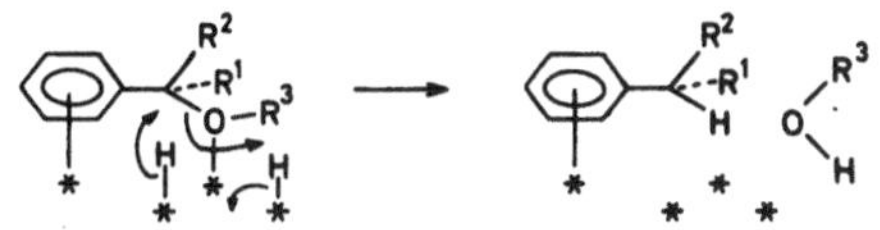

A similar change in the stereochemical course of the reaction is observed for the hydrogenolytic ring opening of phenyloxiranes ($R^1, R^3 = -CH_2-$)[28]. The different behaviour of palladium and nickel has been explained by the higher affinity of the latter towards the oxygen atom, leading to frontside attack of the reducing agent on nickel.

However, a slight modification of R^1 and R^2 may reverse the stereochemical course, as is shown by the following examples[25-29].

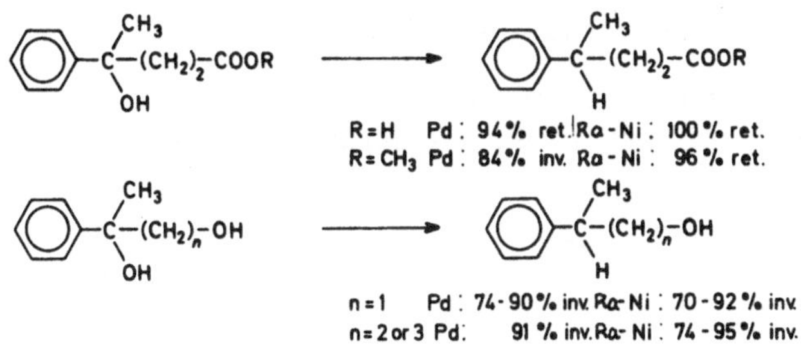

R=H   Pd: 94% ret. Ra-Ni: 100% ret.
R=CH₃ Pd: 84% inv. Ra-Ni: 96% ret.

n=1   Pd: 74-90% inv. Ra-Ni: 70-92% inv.
n=2 or 3 Pd: 91% inv. Ra-Ni: 74-95% inv.

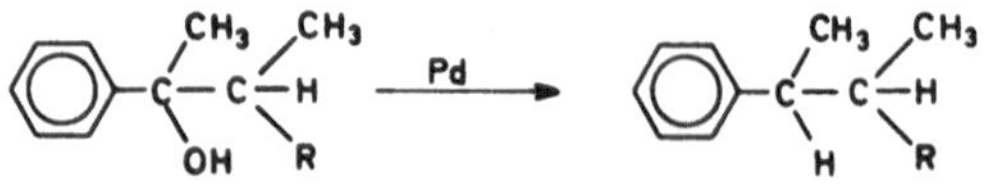

R = H   91 - 95 % inv.
R = CH_3 50 % inv.

Clearly, steric and electronic influences on the interaction of the reactant with the catalyst surface play an important role. More insight is required into the interaction of functional groups with metal surface atoms in order to explain the phenomenon more satisfactorily.

Selectivity

Carbon-carbon double and triple bonds, nitro and cyano groups and carbon-halogen bonds are not resistant towards the conditions required for the (sp^3)carbon-oxygen hydrogenolysis. Oxirane and benzyl-oxygen hydrogenolysis usually leaves carbon-halogen bonds, aliphatic ketonic groups and tetrasubstitued C=C bonds intact. The addition of base (*e.g.* NH_3, NEt_3, KOH) prevents the hydrogenolysis of benzyl alcohols and alkyl benzyl ethers over palladium, whereas the reaction rate for benzyl esters and aryl benzyl ethers is hardly influenced[20].

Application

Although the stereochemistry of the benzyl-oxygen hydrogenolysis cannot be predicted with certainty in all instances, the examples given illustrate the potential applicability of the reaction for the preparation of chiral hydrocarbons from the corresponding chiral alcohol derivatives. The reaction is also extremely useful for the use of benzyl as a protecting group for alcohols, carboxylic acids, and amines[30], in particular in the field of peptide chemistry[31].

$$\text{ROH} + \text{BzlOH} \xrightarrow{H^+} \text{ROBzl} \xrightarrow{H_2/Pd} \text{ROH} + \text{BzlH}$$

$$\text{RCOOH} + \text{BzlOH} \xrightarrow{H^+} \text{RCOOBzl} \xrightarrow{H_2/Pd} \text{RCOOH} + \text{BzlH}$$

Amines are conveniently protected as their benzyloxycarbonamides (12); these may be converted back to the free amines by hydrogenolysis over palladium[32].

$$\text{BzlOC(=O)-Cl} + H_2NR \rightarrow \underset{\mathbf{12}}{\text{BzlOC(=O)-N(H)R}} \xrightarrow{H_2/Pd} \text{BzlH} + [\text{HOC(=O)-N(H)R}] \rightarrow \text{BzlH} + CO_2 + H_2NR$$

The carboxyamide formed first decomposes spontaneously into carbon dioxide and the amine. Another application for this reaction is the peptide synthesis in solution according to the following scheme[33].

$$N\langle\bigcirc\rangle\text{—CH}_2\text{—O—}\overset{\overset{O}{\|}}{C}\text{—}\overset{H}{\underset{}{N}}\text{—}\overset{H}{\underset{}{N}}\text{—}\overset{\overset{O}{\|}}{C}\text{—}\overset{H}{\underset{R}{C}}\text{—NH}_2$$

peptide ↓ synthesis

$$N\langle\bigcirc\rangle\text{—CH}_2\text{—O—}\overset{\overset{O}{\|}}{C}\text{—}\overset{H}{\underset{}{N}}\text{—}\overset{H}{\underset{}{N}}\text{—}\overset{\overset{O}{\|}}{C}\text{—}\overset{H}{\underset{R}{C}}\text{—NH—(peptide)}$$

H_2 ↓ Pd

$$N\langle\bigcirc\rangle\text{—CH}_3 + CO_2 + H_2N\text{—}\overset{H}{\underset{}{N}}\text{—}\overset{\overset{O}{\|}}{C}\text{—}\overset{H}{\underset{R}{C}}\text{—NH—(peptide)}$$

13

The basic pyridine 'handle' readily enables isolation and purification of the coupling product at each stage of the peptide synthesis; final hydrogenolysis of the product yields the hydrazide 13 required for subsequent fragment coupling.

It may be noted that the hydrogenolytic removal of N-benzyloxycarbonyl groups fails with sulphur-containing peptides due to catalyst poisoning (*cf.* p 136). Good yields were obtained for methionine- and S-benzylcysteine-containing peptides when liquid ammonia was used as the solvent and palladium black as the catalyst at -33°[34,35]. Under these conditions benzyl ester and ether C-O bonds were also hydrogenolyzed, whereas S-benzyl groups survived.

Trityl has been frequently used as a selective protecting group for *primary* alcohols (14)[36].

$$RCH_2OH + Ph_3CCl \xrightarrow{\text{pyridine}} RCH_2OCPh_3 \xrightarrow{H_2/Pd} RCH_2OH + HCPh_3$$

14

Phosphoric acids may be protected as their benzyl esters (15), by analogy with the carboxylic acids.

$$(RO)_2\overset{\overset{O}{\|}}{P}\text{-Cl} + BzlOH \xrightarrow{\text{pyridine}} (RO)_2\overset{\overset{O}{\|}}{P}\text{-O-Bzl} \xrightarrow{H_2/Pd} (RO)_2\overset{\overset{O}{\|}}{P}\text{-OH} + HBzl$$

15

For instance, monoalkyl phosphates (17) have been synthesized with the aid of hydrobenzoin (16) as the protective group according to the following route[37] (next page).

The catalytic system $RhCl_3(pyridine)_3$-$NaBH_4$ in dimethylformamide in the presence of hydrogen is able to hydrogenolyze the benzyl-oxygen bond of benzyl acetate[38]. Some doubt arises concerning the homogeneity of the catalyst since no C-O hydrogenolysis occurs in benzyl acetate groups linked to a styrene-divinylbenzene copolymer[39].

123

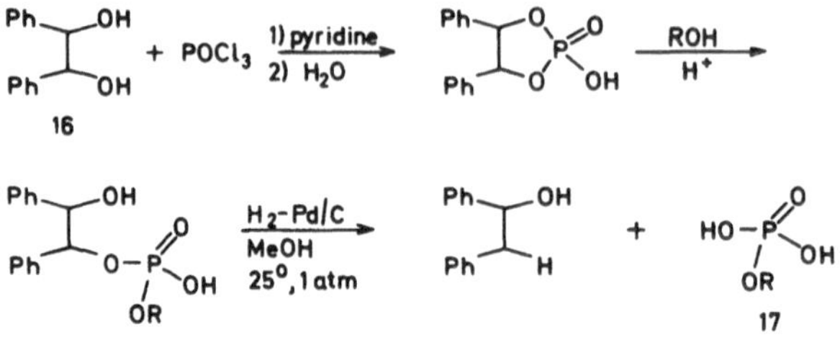

Finally, benzyl-oxygen hydrogenolysis is a valuable means for the direct conversion of aryl ketones to the methylene compounds without the need for the protection of aliphatic ketonic groups present in the molecule.

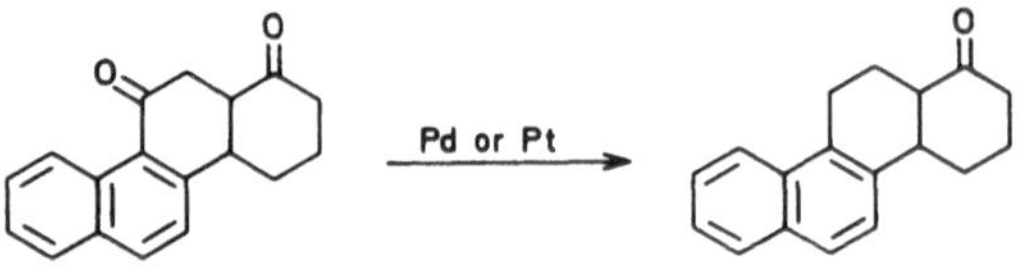

Such a technique is often required in steroid synthesis[40].

(sp²)CARBON-OXYGEN BONDS

Vinyl-oxygen bonds

Hydrogenolysis of vinyl carbon-oxygen bonds is frequently accompanied by hydrogenation of the unsatured function. An example is given by the conversion of the easily accessible 2-benzoyloxymethylene-1-tetralones (18) to the respective 2-methyl-1-tetralones (19) developed for the selective monomethylation of 1-tetralone derivatives[41].

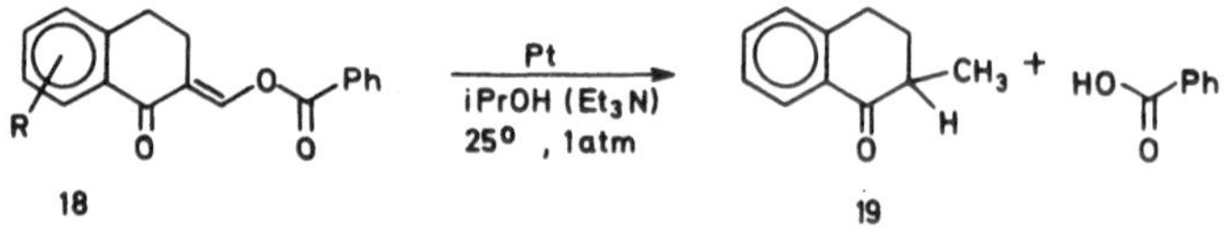

R	Yield (%)
H	89
5-OCH$_3$	91
6-OCH$_3$	97
7-t-C$_4$H$_9$	93
6-NHCOCH$_3$	85

Selective vinyl carbon-oxygen hydrogenolysis with a hydridoruthenium complex has been observed for some vinyl esters[42]:

124

The high activity and selectivity of ruthenium towards the hydrogenolysis reaction is rather surprising.

Aryl-oxygen bonds

An example in this field is the hydrogenolytic removal of the phenyl groups of phenyl phosphates[43].

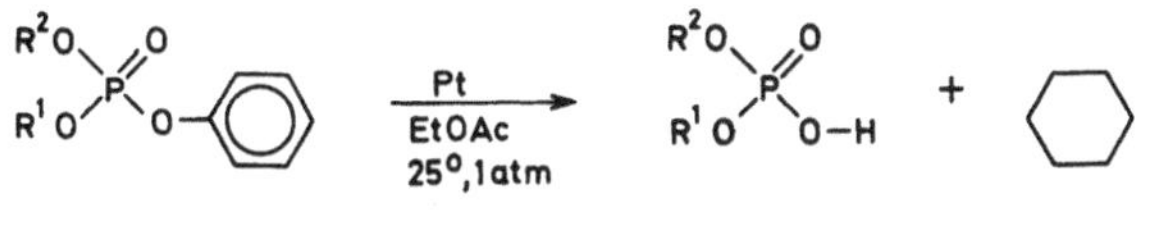

With the uptake of 4 moles of hydrogen, hydrogenolysis of the carbon-oxygen bond occurs prior to or simultaneously with the hydrogenation of the benzene nucleus[44]. Alkyl phosphates are resistant towards hydrogenolysis. The phenyl group has found application as a protective group (in addition to the benzyl group) in the synthesis of sugar phosphates[43].

Several procedures have been developed for the hydrogenolytic cleavage of aryl carbon-oxygen bonds *without affecting the aromatic nucleus*. These methods are entirely based on the good leaving group principle (*cf.* benzyl-oxygen hydrogenolysis).

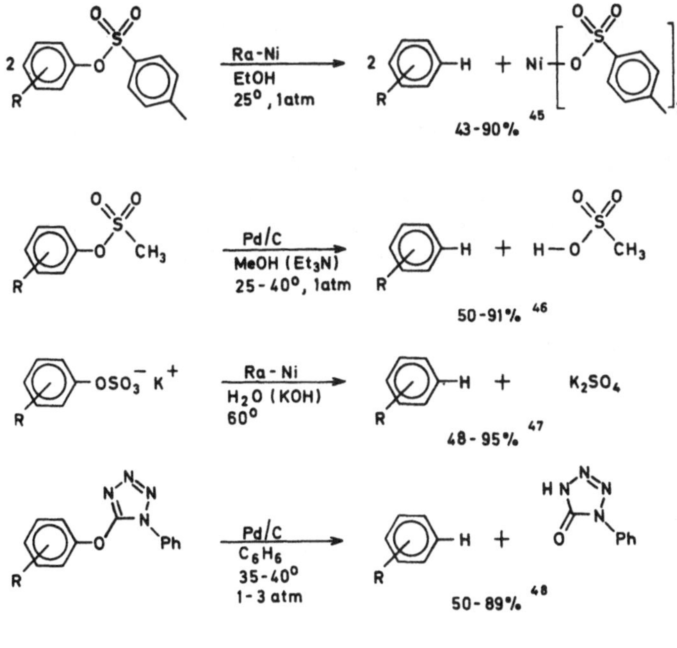

125

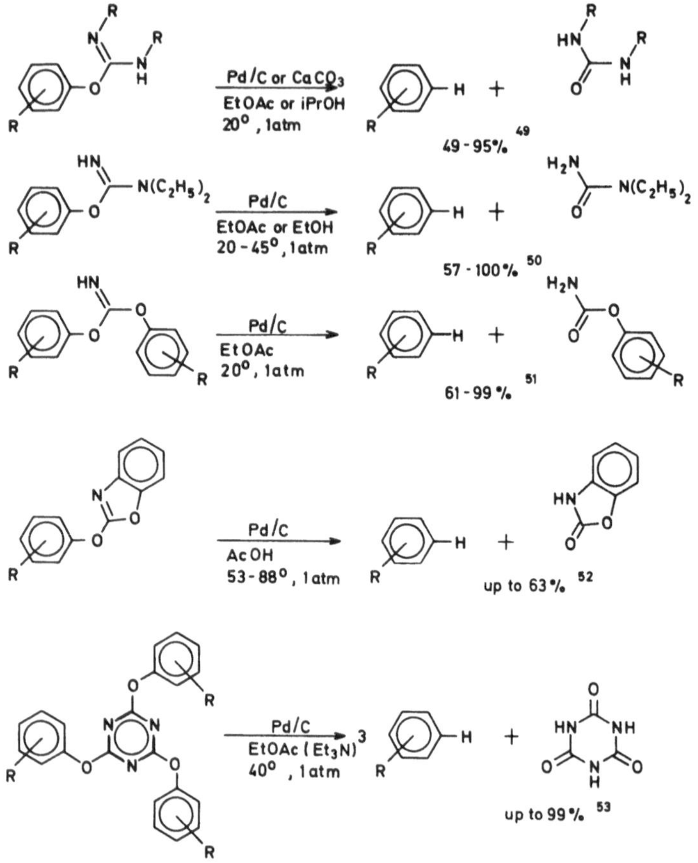

Pd/C or CaCO₃
EtOAc or iPrOH
20°, 1atm

49-95% [49]

Pd/C
EtOAc or EtOH
20-45°, 1atm

57-100% [50]

Pd/C
EtOAc
20°, 1atm

61-99% [51]

Pd/C
AcOH
53-88°, 1atm

up to 63% [52]

Pd/C
EtOAc (Et₃N)
40°, 1atm

up to 99% [53]

The reactants are conveniently prepared from the respective phenols, so that the hydrogenolysis reaction may be considered as a useful method for the replacement of phenolic hydroxyl groups by hydrogen. Easily reducible groups such as nitro, the halogens (except fluorine), and carbon-carbon double and triple bonds are also affected under the reaction conditions required for the carbon-oxygen cleavage.

Examples include the selective removal of one hydroxyl group in certain dihydroxytoluenes[54],

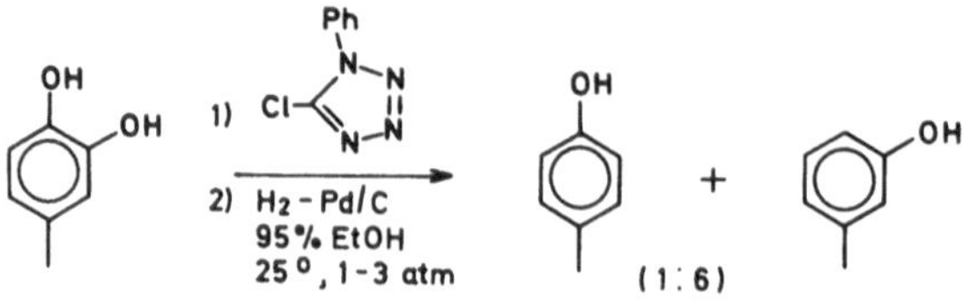

1) Cl—(tetrazole, Ph)
2) H₂ - Pd/C
95% EtOH
25°, 1-3 atm

(1:6)

the tritium labelling of 1,2,3,4-tetrahydronaphthalene[55],

HO——[tetrahydronaphthalene ring]

1) Cl—[Ph-tetrazole]

2) T$_2$ – Pd/CaCO$_3$
EtOH
25° , 1 atm

T——[tetrahydronaphthalene ring]

and the preparation of methyl 3-t-butylbenzoate from methyl salicylate[53]:

COOCH$_3$
OH

$\xrightarrow[H^+]{tBuOH}$

COOCH$_3$
OH
(t-butyl)

$\xrightarrow[\substack{NaOH, H_2O, acetone \\ 25°}]{}$ (cyanuric chloride: Cl—N—Cl triazine, Cl)

H$_3$COOC ... COOCH$_3$ (triazine triether structure with t-butyl groups)
H$_3$COOC

$\xrightarrow[\substack{EtOAc (Et_3N) \\ 40°, 1 atm}]{H_2 – Pd/c}$

3 [COOCH$_3$ aryl with t-butyl]

+ cyanuric acid (O=C–NH ring, HN, NH, O)

The last method needs only cyanuric chloride as the reagent and offers an inexpensive and convenient alternative to the other methods mentioned. Hydrogenolytic cleavage of the three carbon-oxygen bonds occurs during only one residence of the molecule on the catalyst surface, because of the strongly adsorbed triazine system. The addition of a base or the use of a catalyst support is necessary, since the cyanuric acid formed is very strongly adsorbed and poisons the catalyst. The base promotes the transport of cyanuric acid from the catalyst surface to other crystallization sites, whereas the support offers many crystallization sites and takes cyanuric acid from the palladium surface[53].

The hydrogenolyis reaction of the aryl carbon-oxygen bond must be interpreted as a hydrogen atom rather than a hydride attack, contrary to the (sp^3)carbon-oxygen hydrogenolysis. On the other hand a good leaving group seems to be essential for the hydrogenolytic cleavage of the aryl carbon-oxygen bond. In this respect, the hetero(cyclic) ethers exhibit their leaving nature by the preferential formation of carbonyl structures.

[reaction scheme: surface-bound ether → surface-bound aryl-H + carbonyl]

127

Carboxylic acids, esters, and anhydrides

The mild reduction of *2-pyridyl esters* of aromatic carboxylic acids (**20**) over palladium[56] may be interpreted analogously by assuming hydrogenolysis of the carbonyl-oxygen bonds as the first step of the reaction.

20

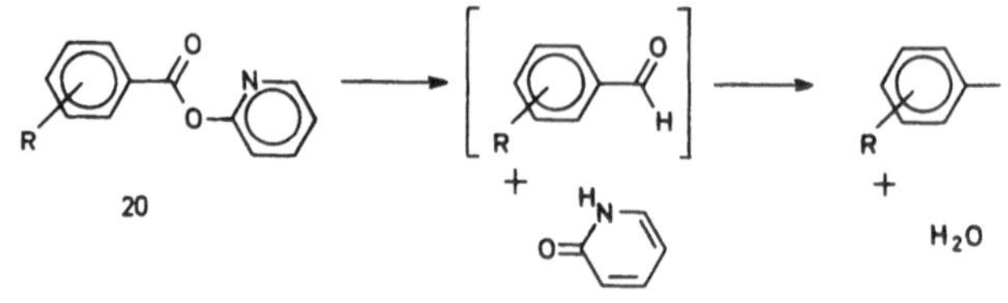

(H_2 - Pd, EtOH or dioxane, 25-55°, 1-3 atm.)

The benzaldehyde formed is then further reduced via the benzyl alcohol to the corresponding toluene.

In general, *other esters* of carboxylic acids in addition to the *acids* themselves require rather drastic reaction conditions (150–320°, 100–300 atm)[57–60]. Some examples are given below.

$$C_2H_5OOC(CH_2)_4COOC_2H_5 \xrightarrow[255°,\ 130-200\,atm]{CuCr_2O_4} HOCH_2(CH_2)_4CH_2OH + 2\ C_2H_5OH$$

85-90 %[57]

$$CH_3(CH_2)_7CH{=}CH(CH_2)_7COOC_4H_9 \xrightarrow[300°,\ 200\,atm]{ZnCr_2O_4} CH_3(CH_2)_7CH{=}CH(CH_2)_7CH_2OH + C_4H_9OH$$

65 %[58]

($CuCr_2O_4$ gives the saturated alcohol)

$$HOOC(CH_2)_3COOH \xrightarrow[H_2O,\ 250°,\ 179\,atm]{Re_2O_7} HOCH_2(CH_2)_3CH_2OH + 2\ H_2O$$

100 %[59]

Here, appreciable amounts of esters may be obtained from the reaction between the acid and the alcohol formed, *e.g.*[59]

$$CH_3(CH_2)_3COOH \xrightarrow[160°,\ 173\,atm]{Re_2O_7} CH_3(CH_2)_3CH_2OH\ \ 55\% \quad + \quad CH_3(CH_2)_3COO(CH_2)_4CH_3\ \ 43\%$$

In some cases, it is possible to reduce the carboxylic acid group without concomitant saturation of a C=C bond[61].

$$H_2C{=}CH(CH_2)_7COOH \xrightarrow[undecane,\ 250°,\ 180\,atm]{Cu(II) + Cd(II)} H_2C{=}CH(CH_2)_7CH_2OH$$

up to 90 %

The application of large amounts of catalyst as well as extended reaction times results in the hydrogenolysis of *cyclic anhydrides* under mild conditions[62].

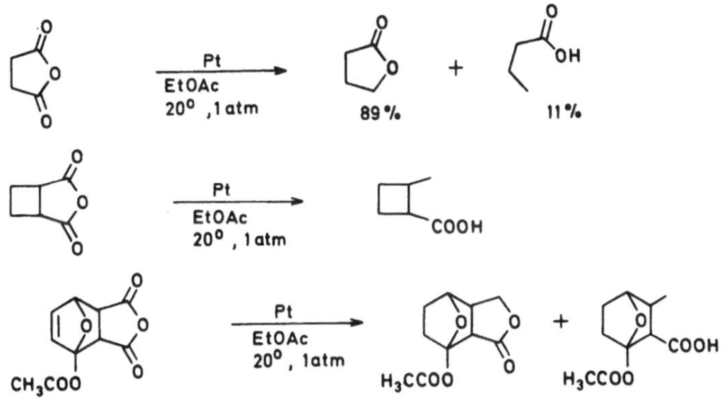

The examples are given merely in order to demonstrate that the application of long reaction times may cause (undesirable) reduction of anhydride groups in the reactant. Recently, it has been shown that $RuCl_2(Ph_3P)_3$ is an active catalyst for the homogeneous hydrogenolysis of carboxylic acid anhydrides[63]:

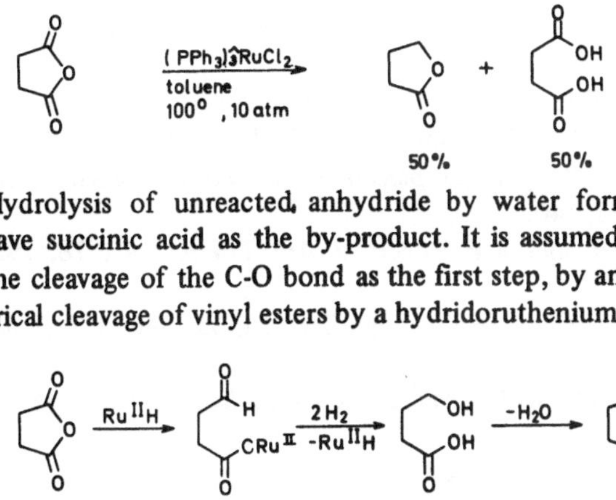

Hydrolysis of unreacted anhydride by water formed during the reduction gave succinic acid as the by-product. It is assumed that the reaction involves the cleavage of the C-O bond as the first step, by analogy with the stoichiometrical cleavage of vinyl esters by a hydridoruthenium complex[42] (see p 125).

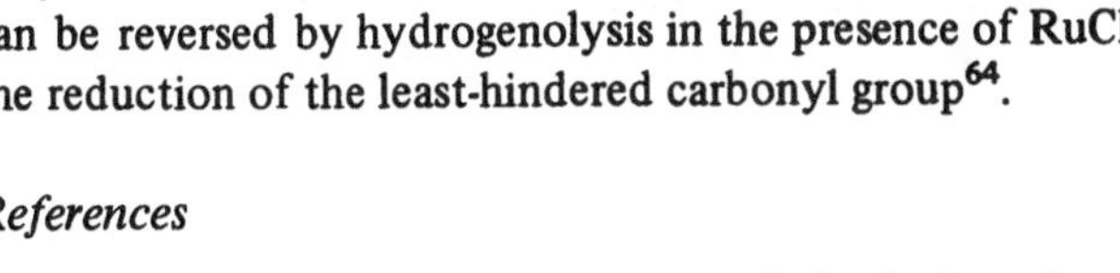

Finally, it may be noted that the highly regioselective partial reduction of unsymmetrical cyclic anhydrides to the corresponding γ-lactones with $LiAlH_4$ can be reversed by hydrogenolysis in the presence of $RuCl_2(Ph_3P)_3$, resulting in the reduction of the least-hindered carbonyl group[64].

References

1. G.W. Brown, in *The Chemistry of the hydroxyl group*, Part 1, S. Patai, Ed., Interscience Publ., London, 1971, p 626.
2. A. Holt, A.W.P. Jarvie, and J.J. Mallabar, *J. Organomet. Chem.*, **59**, 141 (1973).

129

3. E. Vowinkel and I. Büthe, *Chem. Ber.*, **107**, 1353 (1974).
4. E.W. Garbisch, Jr., *Chem. Commun.*, **1967**, 806.
5. W.L. Howard and J.H. Brown, Jr., *J. Org. Chem.*, **26**, 1026 (1961).
6. M. Verzele, M. Acke, and M. Anteunis, *J. Chem. Soc.*, **1963**, 5598.
7. W.S. Emerson, *Org. React.*, **4**, 174 (1948).
8. M.L. Moore, *Org. React.*, **5**, 301 (1949).
9. J. Volf, J. Pasek, and P. Kraus, *Sci. Pap. Inst. Chem. Technol. Prague C*, 19(5), 27 (1973), *Chem. Abstr.*, **79**, 125724*b* (1973).
10. J.A. Davies, C.H. Hassall, and I.H. Rogers, *J. Chem. Soc. C*, **1969**, 1358.
11. D.E. Pearson and J.D. Bruton, *J. Am. Chem. Soc.*, **73**, 864 (1951).
12. F.S. Dovell and H. Greenfield, *J. Org. Chem.*, **29**, 1265 (1964).
13. F.S. Dovell and H. Greenfield, *J. Am. Chem. Soc.*, **87**, 2767 (1965).
14. H. Greenfield, *Ann. N.Y. Acad. Sci.*, **145**, 108 (1967).
15. G.V. Pigulevskii and Z.Y. Rubashko, *Zh. Obshch. Khim.*, **25**, 2227 (1955), *Chem. Abstr.*, **50**, 9291*a* (1956).
16. F.J. McQuillin and W.O. Ord, *J. Chem. Soc.*, **1959**, 3169.
17. G.J. Park and R. Fuchs, *J. Org. Chem.*, **22**, 93 (1957).
18. R.W. Meschke and W.H. Hartung, *J. Org. Chem.*, **25**, 137 (1960).
19. P.N. Rylander and D.R. Steele, *Engelhard Ind. Tech. Bull.*, **6** (2), 41 (1965).
20. A.P.G. Kieboom, J.F. de Kreuk, and H. van Bekkum, *J. Catal.*, **20**, 58 (1971).
21. A.M. Khan, F.J. McQuillin, and I. Jardine, *J. Chem. Soc. C*, **1967**, 136.
22. H. Messchendorp and A.P.G. Kieboom, unpublished results.
23. A. Streitwieser, Jr., *Solvolytic Displacement Reactions*, McGraw-Hill, New York, 1962, pp 95, 140.
24. F.C. Uhle, J.E. Krueger, and A.E. Rogers, *J. Am. Chem. Soc.*, **78**, 1932 (1956).
25. E.I. Klabunovskii, *Russ. Chem. Rev.* (Engl. Transl.), **35**, 546 (1966), and references.
26. S. Mitsui, Y. Kudo, and M. Kobayashi, *Tetrahedron*, **25**, 1921 (1969), and references.
27. S. Mitsui, S. Imaizumi, and Y. Esashi, *Bull. Chem. Soc. Jpn.*, **43**, 2143 (1970), and references.
28. S. Mitsui and Y. Nagahisa, *Chem. Ind. (London)*, **1965**, 1975.
29. A.P.G. Kieboom, *React. Kinet. Catal. Lett.*, **1**, 433 (1974).
30. W.H. Hartung and R. Simonoff, *Org. React.*, **7**, 263 (1953).
31. M. Bodanszky and M.A. Ondetti, *Peptide Synthesis*, Interscience Publ., New York, 1966, p 21 ff.
32. J.W. Barton, in *Protective Groups in Organic Chemistry*, J.F.W. McOmie, Ed., Plenum Press, New York, 1973, p 56.
33. R. Macrae and G.T. Young, *J. Chem. Soc., Chem. Commun.*, **1974**, 446.
34. J. Meienhofer and K. Kuromizu, *Tetrahedron Lett.*, **1974**, 3259.
35. K. Kuromizu and J. Meienhofer, *J. Am. Chem. Soc.*, **96**, 4978 (1974).

36. *Cf.* J.-Y. Rouault, C. Morpain, M. Tisserand, and E. Cerutti, *C.R. Acad. Sc., Ser. C,* 277, 787 (1973).
37. T. Ukita, K. Nagasawa, and M. Irie, *J. Am. Chem. Soc.,* 80, 1373 (1958).
38. C.J. Love and F.J. McQuillin, *J. Chem. Soc., Perkin Trans. 1,* 1973, 2509.
39. A.P.G. Kieboom, unpublished result.
40. D. Taub, in *The Total Synthesis of Natural Products,* Vol. 2, J. ApSimon, Ed., Wiley, New York, 1973, pp 657, 660.
41. A.P.G. Kieboom and H. van Bekkum, *Synthesis,* 1970, 476.
42. S. Komiya and A. Yamamoto, *J. Chem. Soc., Chem. Commun.,* 1974, 523.
43. F.J. Reithel and C.K. Claycomb, *J. Am. Chem. Soc.,* 71, 3669 (1949).
44. A. Jung and R. Engel, *J. Org. Chem.,* 40, 244 (1975).
45. W. Kenner and M.A. Murray, *J. Chem. Soc.,* 1949, S 178.
46. K. Clauss and H. Jensen, *Angew. Chem.,* 85, 981 (1973).
47. W. Lonsky, H. Traitler, and K. Kratzl, *J. Chem. Soc., Perkin Trans. 1,* 1975, 169.
48. W.J. Musliner and J.W. Gates, Jr., *J. Am. Chem.* Soc., 88, 4271 (1966); W.J. Musliner and J.W. Gates, Jr., *Org. Synth.,* 51, 82 (1971).
49. E. Vowinkel and C. Wolff, *Chem. Ber.,* 107, 907 (1974).
50. E. Vowinkel and H.J. Baese, *Chem. Ber.,* 107, 1213 (1974).
51. E. Vowinkel and C. Wolff, *Chem. Ber.,* 107, 1739 (1974).
52. E.J. Eisenbraun, J.W. Burnham, J.D. Weaver, and T.E. Webb, *Ann. N.Y. Acad. Sci.,* 214, 204 (1973); J.D. Weaver, E.J. Eisenbraun, and L.E. Harris, *Chem. Ind. (London),* 1973, 187.
53. A.W. van Muijlwijk, A.P.G. Kieboom, and H. van Bekkum, *Rec. Trav. Chim. Pays-Bas,* 93, 204 (1974).
54. C.F. Barfknecht, R.V. Smith, and V.D. Reif, *Can. J. Chem.,* 48, 2138 (1970).
55. P.J. van der Jagt, W. den Hollander, and B. van Zanten, *Tetrahedron,* 27, 1049 (1971).
56. C.J. Cavallito and T.H. Haskell, *J. Am. Chem. Soc.,* 66, 1166 (1944).
57. W.A. Lazier, J.W. Hill, and W.J. Amend, *Org. Synth.,* Coll. Vol. 2, 325 (1947).
58. J. Sauer and H. Adkins, *J. Am. Chem. Soc.,* 59, 1 (1937).
59. H.S. Broadbent, G.C. Campbell, W.J. Bartley, and J.H. Johnson, *J. Org. Chem.,* 24, 1847 (1959).
60. K.M.K. Muttzall, *High-pressure hydrogenation of fatty acid esters to fatty alcohols,* Ph.D. Thesis, Delft University of Technology, Delft, 1966.
61. J. Perdijk, *Koperhydride complexen en hun hydrogenenerende eigenschappen,* Ph.D. Thesis, Twente University of Technology, Enschede, 1969.
62. R. McCrindle, K.H. Overton, and R.A. Raphael, *Proc. Chem. Soc. (London),* 1961, 313 and *J. Chem. Soc.,* 1962, 4798.
63. J.E. Lyons, *J. Chem. Soc., Chem. Commun.,* 1975, 412.
64. P. Morand and M. Kayser, *Ibid.,* 1976, 314.

5. Carbon-nitrogen hydrogenolysis

Hydrogenolysis of the carbon-nitrogen bond closely resembles that of the carbon-oxygen bond, but takes place less readily. As expected, hydrogenolytic ring opening of aziridines occurs under very mild conditions (*cf.* the oxiranes), whereas benzylamine derivatives are hydrogenolyzed at a reasonable rate[1] under the conditions given for benzyl alcohol compounds (p 118). Although palladium is most often the catalyst of choice for carbon-nitrogen hydrogenolysis, both platinum and Raney nickel have been appleid successfully. The selectivity towards other reducible groups in the molecule is similar to that of the carbon-oxygen hydrogenolysis reaction (p 122).

MECHANISM

In accordance with the proton catalysis found for benzyl oxygen derivatives, the addition of strong acids or the use of quaternary nitrogen compounds enhances the reaction rate considerably by the formation of a better leaving group. The reaction may be considered as a nucleophilic displacement according to

in which a proton leaves the surface by way of the nitrogen atom. In this way, even cleavage of the aliphatic carbon-nitrogen bond may occur in some unusual cases, *e.g.*[2]

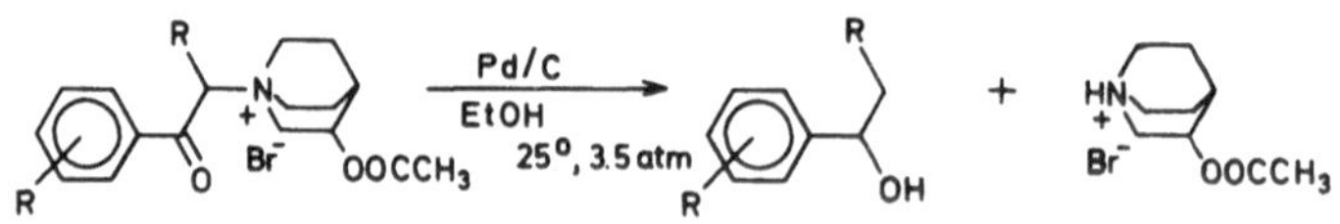

64 - 91%

The stereochemistry of the carbon-nitrogen hydrogenolysis reaction is different from that of the carbon-oxygen hydrogenolysis. Here, both Raney nickel and palladium show inversion of configuration at the benzylic carbon for several benzylamine derivatives[3-5].

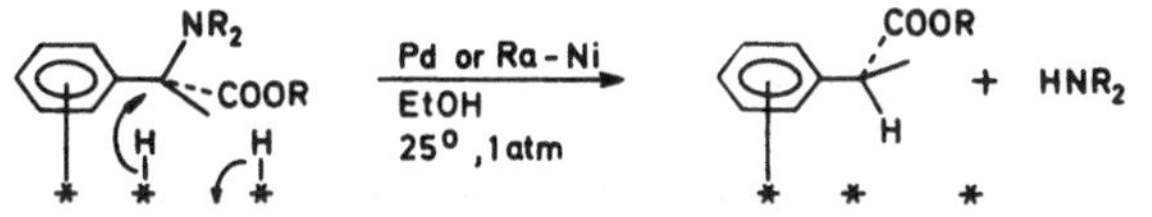

R = H, Me, Et : up to 99% inversion

Hydrogenolysis of 2-phenylaziridines (1) also takes place with inversion of configuration at the benzylic carbon if palladium is the catalyst. Raney nickel shows a slight preference towards retention, whereas this preference is much higher with platinum as the catalyst[6,7].

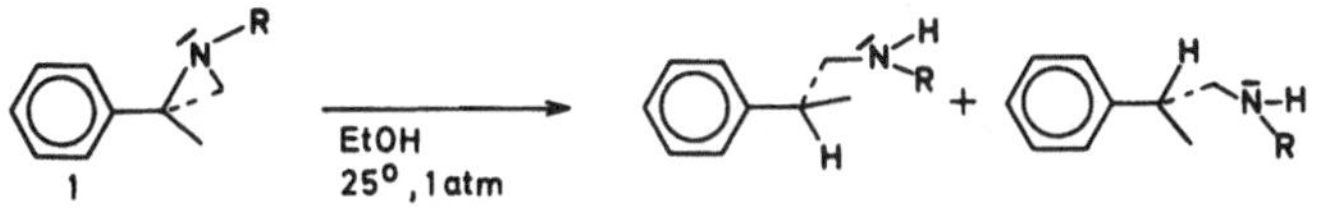

R = H, CH$_3$, COCH$_3$

catalyst	inversion (%)	retention (%)
Pd	86 - 88	12 - 14
R$_\alpha$ Ni	38 - 44	56 - 62
Pt	3 - 24	76 - 97

It is assumed that different electronic interactions between the lone pair of the nitrogen and the various metals will be largely responsible for these phenomena. In this respect, the alteration of the stereochemical course of the hydrogenolysis of 2-methyl-2-phenylaziridine (1, R=H) over palladium upon addition of increasing amounts of strong base is noteworthy[7]:

NaOH (equivalents)	Inversion (%)	Retention (%)
0.00	86	14
0.01	73	27
0.02	21	79
0.05	10	90
0.10	0	100

This effect is not observed for the corresponding N-methyl and N-acetyl substituted derivatives. Evidently ionization of the N-H bond plays a major role in the change of stereochemistry of the reaction. Since hydroxyl anions exhibit good affinity for the palladium surface, the following equilibrium will occur on the catalyst metal:

133

in which the right-hand adsorbed state favours the hydrogen attack by the $S_{N}i$ mechanism (retention of configuation).

The influence of ionic species on the stereochemical course of the carbon-nitrogen hydrogenolysis is further demonstrated by the hydrogenolysis of different quaternary benzylammonium salts over palladium[5].

	X⁻	Retention (%)	Inversion (%)
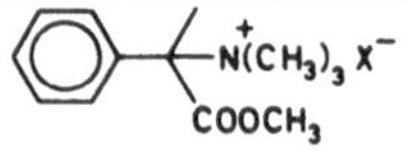	I⁻	50	50
	Br⁻	30	70
	AcO⁻	17	83
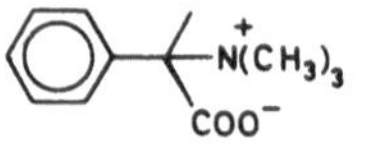	–	9	91

Pd/C or Pd/BaSO₄, EtOH (Et₃N), 25°, 1 atm

The order of affinity between the various anions and the palladium surface according to the hard and soft acid/base criterion[8] is expected to be: $I^{-} > Br^{-} > AcO^{-}, -COO^{-}$[,5]. The following equilibrium is proposed for the adsorbed state:

Greater affinity of the anions for the palladium metal will shift this equilibrium to the right, *i.e.* hydrogen attack according to the $S_{N}i$ mechanism becomes more accessible in addition to the $S_{N}2$ mode of reaction. In this manner, the lower stereoselectivity in the case of the halogens can be understood.

APPLICATION

The benzyl-nitrogen hydrogenolysis has been applied to the temporary protection and to the synthesis of *tertiary* amines via their benzylammonium salts, since the benzyl group can be removed very easily, *e.g.* $2 \rightarrow 3$[9].

Another application is reported[10] for the conversion of aryl carbonyl into aryl methylene compounds via the oxime, semicarbazone or hydrazone as an

134

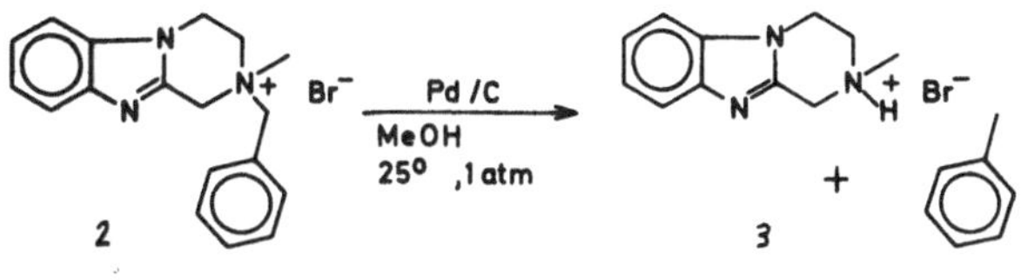

2 ⟶ **3** +

alternative to the direct reduction of the aryl ketonic function (p 124). In order to bring the aryl ketones to the required purity, it is sometimes necessary to convert them to these (crystalline) derivatives[11]. Direct hydrogenolysis of these derivatives affords the respective methylene compounds in excellent yields[9].

R = alkyl, aryl
X = NOH, NNHaryl, NNHCONH$_2$

The copper catalyzed reduction of amides according to

may be considered as an example of (sp^2)carbon-nitrogen hydrogenolysis, assuming the hydrogenolysis of the carbon-nitrogen bond as the first step of the reaction.

References

1. W.H. Hartung and R. Simonoff, *Org. React.,* 7, 263 (1953).
2. R.J. Adamski, R.E. Hackney, S. Numajiri, L.J. Spears, and E.H. Yen, *Synthesis,* **1973**, 221.
3. C. O'Murchu, *Tetrahedron Lett.,* **1969**, 3231.
4. H. Dahn, J.A. Garbarino, and C. O'Murchu, *Helv. Chim. Acta,* 53, 1370, 1379 (1970).
5. Y. Sugi and S. Mitsui, *Tetrahedron,* 29, 2041 (1973), and references.
6. Y. Sugi and S. Mitsui, *Bull. Chem. Soc. Jpn.,* 42, 2984 (1969).
7. Y. Sugi and S. Mitsui, *Bull. Chem. Soc. Jpn.,* 43, 1489 (1970).
8. R.G. Pearson and J. Songstad, *J. Am. Chem. Soc.,* 89, 1827 (1967).
9. J. Schmutz and F. Künzle, *Helv. Chim. Acta,* 39, 1144 (1956).
10. J.W. Burnham and E.J. Eisenbraun, *J. Org. Chem.,* 36, 737 (1971).
11. *Cf.* H. van Bekkum, A.P.G. Kieboom, and K.J.G. van de Putte, *Rec. Trav. Chim. Pays-Bas,* 88, 52 (1969).

6. Carbon-sulphur hydrogenolysis

The low-valent sulphur atom as present in e.g. thiols, sulphides and sulphoxides, has a particular place in the hydrogenation and hydrogenolysis reactions, because of its very high affinity towards transition metals. The strong bond-formation by electron donation from the lone pair(s) of the sulphur to the vacant d-orbitals of the metal may result in serious poisoning of the catalysts[1].

It was early recognized that the toxicity of aliphatic sulphides and thiols towards the hydrogenation of olefins is enhanced by an increase in the chain length of the sulphur compound[2]. The aliphatic chain rotates and moves around, thus obstructing access to a number of adjacent catalyst atoms. The obstructed area estimated from molecular models is roughly in agreement with the relative toxicities found[1]. Reduction of other functional groups of compounds containing low-valent sulphur atoms is frequently inhibited by the strong chemisorption of the sulphur part of the molecule. In some instances, close contact between sulphur and the catalyst may occur less readily for steric reasons or the adsorption of the sulphur atom may force the functional group to be reduced towards the catalyst surface, e.g.[3]

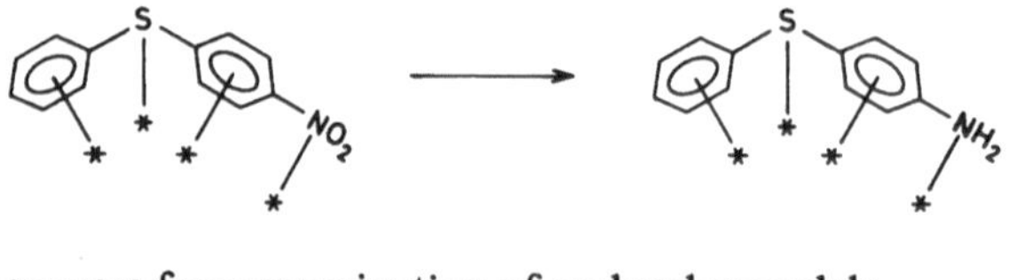

as appears from examination of molecular models.

Hydrogenolysis of the carbon-sulphur bond itself stops at a low stage of conversion.

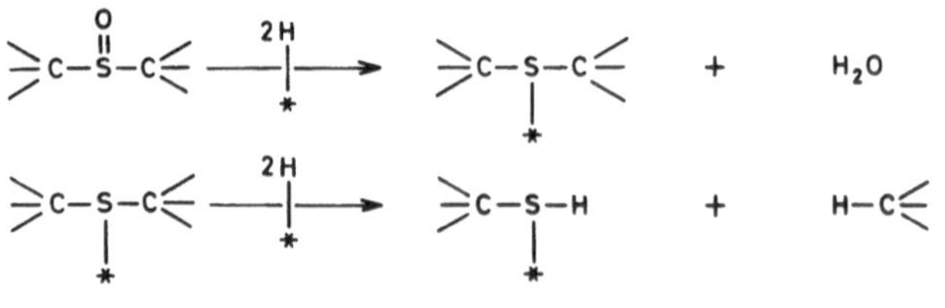

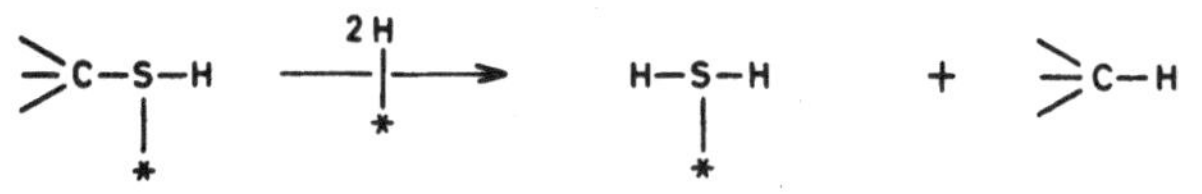

The sulphur-containing products do not desorb from the catalyst surface since the adsorption strengths increase in the order:

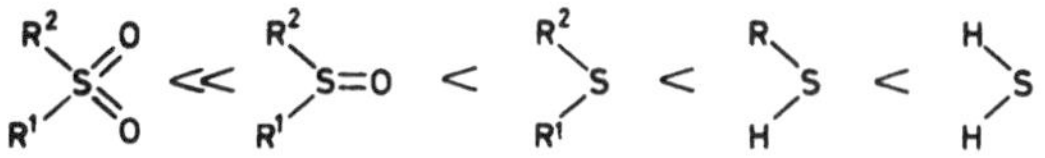

Therefore very large amounts of catalyst are required to achieve complete conversion. This explains the use of a relatively inexpensive catalyst for the carbon-sulphur hydrogenolysis reaction (also denoted as hydrodesulphurization) such as Raney nickel[4]. In addition, the sulphides of cobalt, molybdenum, nickel and some other metals[5,6] have been applied. Raney nickel is preferably used for laboratory purposes, because of the rather vigorous conditions required with sulphide catalysts[6].

REACTIVITY

Thiols, alkyl thioethers, and dithioacetals are readily hydrogenolyzed using large amounts (10-20 times the weight of reactant) of Raney nickel. The amount of hydrogen already chemisorbed on the Raney nickel is generally a sufficient hydrogen source for the reaction. Thiols are more strongly adsorbed and more rapidly hydrogenolyzed than are thioethers; this leads to the selective cleavage of the thiol carbon-sulphur bond[7].

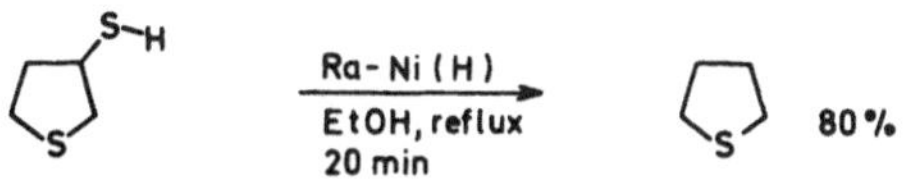

The same holds for the dithioacetals (1) which are conveniently converted to the respective methylene compounds

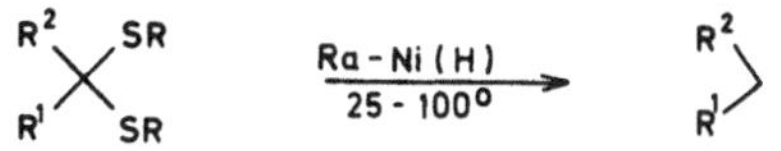

presumably during just one residence on the catalyst surface. Aryl thioethers, sulphoxides and sulphones are hydrogenolyzed less readily[8].

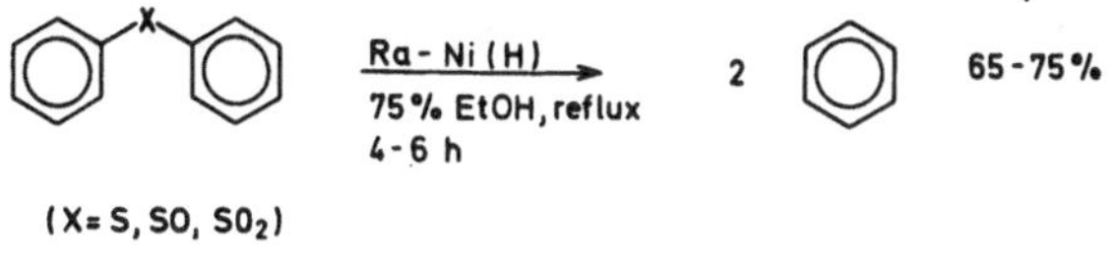

(X= S, SO, SO₂)

Using nickel boride as the catalyst[9], diphenyl sulphide and diphenyl sulphoxide were hydrogenolyzed, whereas diphenyl sulphone remained intact.

It may be noted, that the use of hydrogen-deficient catalysts without supplying hydrogen may yield reductive coupling products[4], *e.g.*[10]

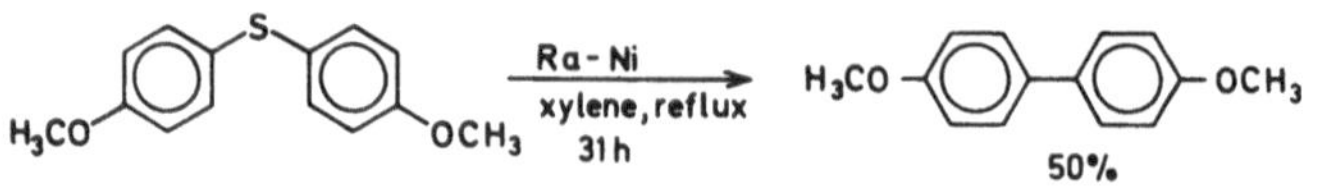

STEREOCHEMISTRY

The stereochemistry of the carbon-sulphur hydrogenolysis has been established for a number of benzyl-sulphur compounds (2)[11-13].

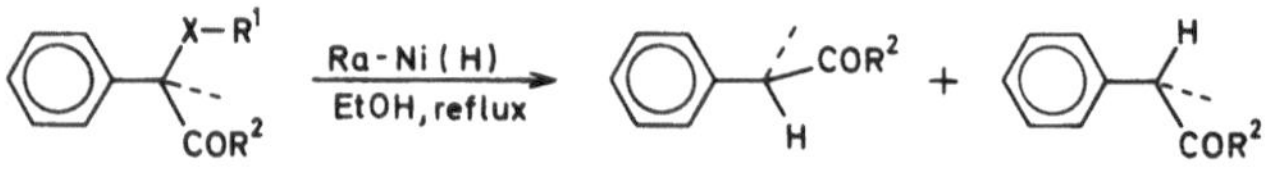

X	inversion (%)	retention (%)
S	50	50
SO	50	50
SO$_2$	75 - 90	10 - 25

The chemisorption of the sulphide and sulphoxide group on the catalyst surface is responsible for the occurence of both front- and backside attack of hydrogen (S$_N$i and S$_N$2 mechanism, respectively), by analogy with the anion effect on the stereochemical course of the hydrogenolysis reaction of benzyl-trimethylammonium derivatives over palladium (*cf.* p 134).

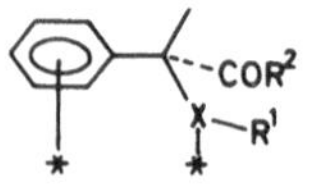

On the other hand, the corresponding benzyl sulphone (3), which contains no low-valent sulphur, will be preferentially adsorbed in the following manner:

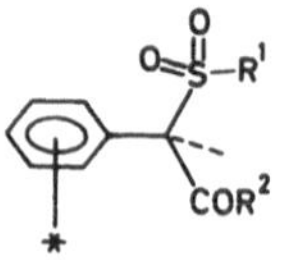

3

resulting in hydrogen attack with inversion of configuration at the benzylic carbon.

SELECTIVITY

Rather high selectivity of carbon-sulphur hydrogenolysis — with respect to other reducible functional groups — has been observed[4] (except in the special cases mentioned above). This selectivity is readily understood in terms of the complete occupation of the catalyst surface (i.e. poisoning for other reducible groups) by the sulphur part of the molecules.

APPLICATION

The hydrogenolysis of dithioacetals is of great value for the selective reduction of ketones and aldehydes according to

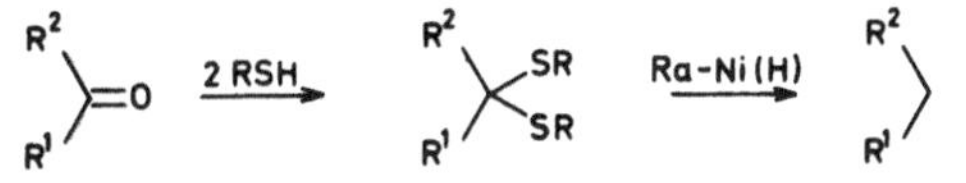

Two examples are given below:

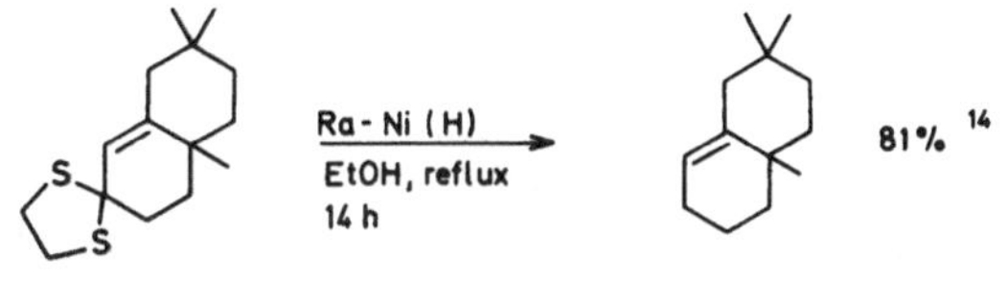

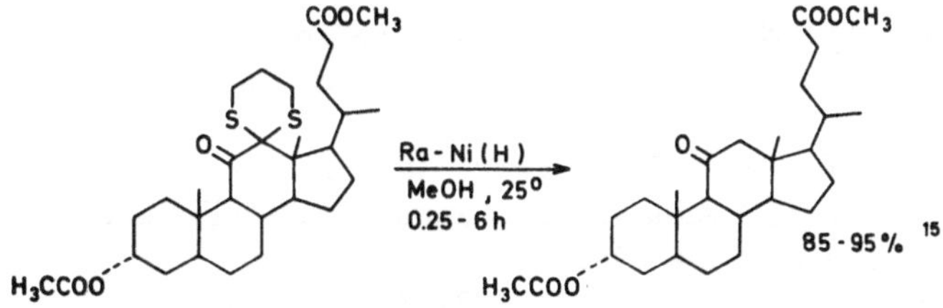

The use of deuterated Raney nickel catalyst leads to the formation of geminal dideutero compounds, e.g.[16]

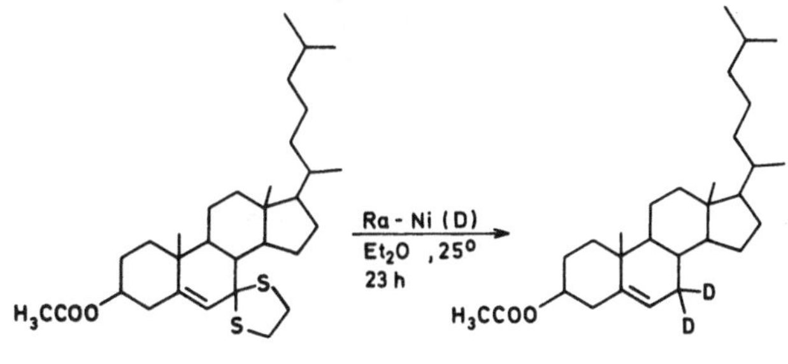

The hydrogenolysis of alkyl thioethers has been applied to the synthesis of t-butyl compounds according to the following scheme[17].

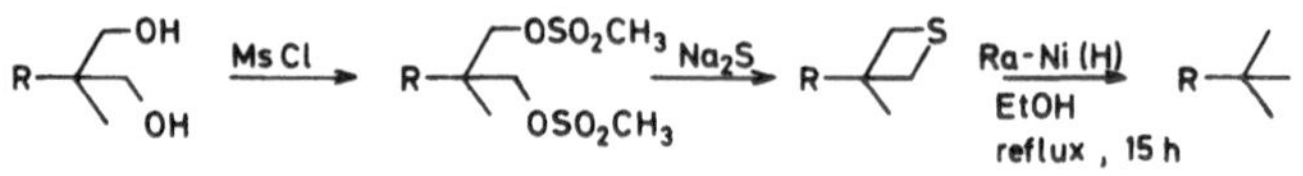

The starting 1,3-diol is easily accessible by the malonic ester synthesis. In this way 1,2-dineopentyl-3,4,5,6-tetramethylbenzene (5) was prepared from the corresponding 1,2-bis(chloromethyl)compound (4)[17].

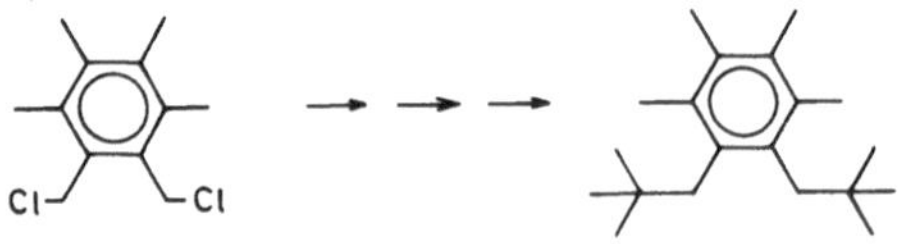

References

1. R. Baltzly, *Ann. N.Y. Acad. Sci.*, **145**, 31 (1967).
2. E.B. Maxted and H.C. Evans, *J. Chem. Soc.*, **1937**, 1004.
3. M. Freifelder, *Ann. N.Y. Acad. Sci.*, **145**, 5 (1967).
4. G.R. Pettit and E.E. van Tamelen, *Org. React.*, **12**, 356 (1962).
5. S.C. Schuman and R. Shalit, *Catal. Rev.*, **4**, 245 (1970).
6. O. Weisser and S. Landa, *Sulphide catalysts, their properties and applications*, Pergamon Press, Oxford, 1973, p 182 ff.
7. L.W.C. Miles and L.N. Owen, *J. Chem. Soc.*, **1952**, 817.
8. R. Mozingo, D.E. Wolf, S.A. Harris, and F. Folkers, *J. Am. Chem. Soc.*, **65**, 1013 (1943).
9. W.E. Truce and F.M. Perry, *J. Org. Chem.*, **30**, 1316 (1965).
10. H. Hauptmann, B. Wladislaw, L.L. Nazario, and W.F. Walter, *Justus Liebigs Ann. Chem.*, **576**, 45 (1952).
11. W.A. Bonner, *J. Am. Chem. Soc.*, **74**, 1034, 5089 (1952).
12. S. Imaizumi, *Nippon Kagaku Zasshi*, **78**, 1396 (1957) and **81**, 633 (1960), *Chem. Abstr.*, **54**, 1403h (1960) and **56**, 402g (1962).
13. E.I. Klabunovskii, *Russ. Chem. Rev.* (Engl. Transl.), **35**, 546 (1966).
14. F. Sondheimer and S. Wolfe, *Can. J. Chem.*, **37**, 1870 (1959).
15. S. Archer, T.R. Lewis, C.M. Martini, and M. Jackman, *J. Am. Chem. Soc.*, **76**, 4915 (1954).
16. D.K. Fukushima, S. Lieberman, and B. Praetz, *J. Am. Chem. Soc.*, **72**, 5205 (1950).
17. M.S. Newman, J.R. LeBlanc, H.A. Karnes, and G. Axelrad, *J. Am. Chem. Soc.*, **86**, 868 (1964).

7. Carbon-halogen hydrogenolysis

Hydrogenolysis of the carbon-halogen bond is applied in the replacement of halogens by hydrogen, for example in the case of the use of the halogen as a protective or directing group during the synthesis of aromatic compounds, in the specific deuterium labelling of organic compounds or in the selective conversion of carboxylic acids, *via* the acid chloride, to the corresponding aldehydes (Rosenmund reduction). In general, the use of palladium as the catalyst in the presence of a base (amine, alkali metal hydroxide or acetate) at ambient temperature and pressure has been found to be most convenient, in particular in polar media. In addition, Raney nickel has been applied with success.

REACTIVITY

As would be expected from solvolytic displacement and metal (or boro) hydride reactions, the rate of hydrogenolysis decreases in the order $I > Br > Cl > F$. The order of reactivity with respect to the carbon atom is somewhat different from that for the above-mentioned homogeneous reactions, *viz*.

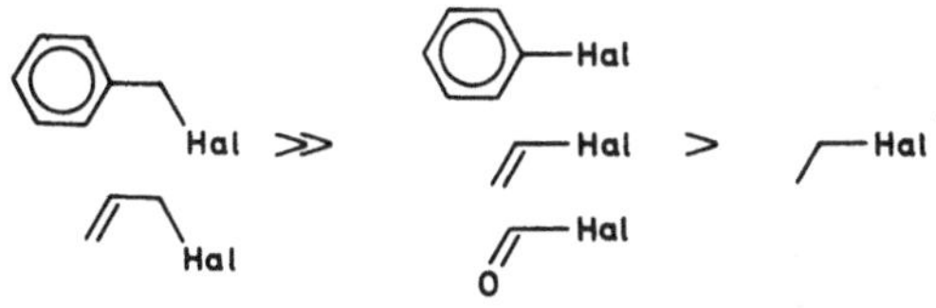

For benzyl and allyl halogen compounds the energy of the transition state will be lowered because of the electronic interaction of the adjacent unsaturated group with the catalyst and the resonance stabilization of the partially positively charged carbon atom.

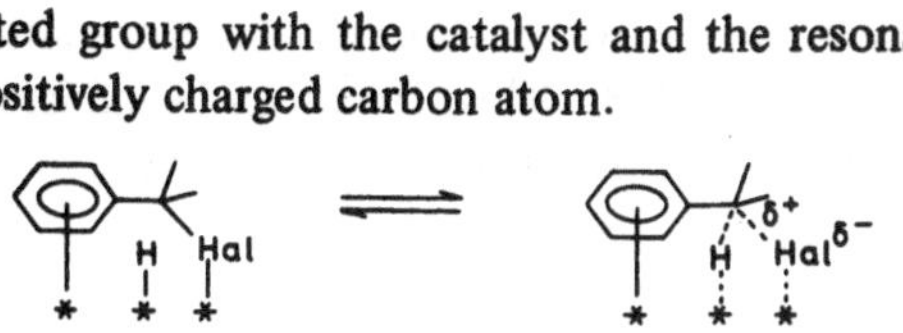

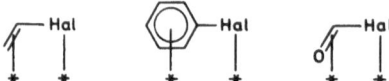

This rate-enhancing effect is quite similar to that in benzyl-oxygen and benzyl-nitrogen hydrogenolysis. In this respect, the greater rate of reaction of aryl halogenides as compared with alkyl halogenides is due to the electronic interaction of the extended π-system with the catalyst metal according to

In this way, the catalytic action towards cleavage of the carbon-halogen bond is enhanced, since this bond is orientated parallel to the catalyst surface. For alkyl halides, the carbon-halogen bond may be expected to lie at a considerable angle to the catalyst surface:

because of steric repulsion between the alkyl chain and the catalyst surface.

The hydrogenolysis of benzyl and allyl carbon-halogen bonds as well as that of alkyl and aryl iodides and *tertiary* alkyl chlorides and bromides occurs very easily. The other alkyl halides require a basic polar medium (*e.g.* ethanolic KOH) with palladium or Raney nickel as the catalyst at ambient conditions. Alkyl fluorides are reasonably stable under these conditions, whereas aryl fluorides are hydrogenolyzed slowly. Carbonyl, vinyl and aryl halides (Cl, Br, I) are smoothly hydrogenolyzed in the presence of acid acceptors.

The hydrogen halide formed during the reaction has a high affinity for the catalyst metal and competes with the organic halogen compound on the catalyst surface.

Increasing amounts of hydrogen halide shift the adsorption equilibrium to the right, resulting in a progressively declining rate of reaction. The hydrogenolysis is accelerated considerably by the addition of base. This effect may be explained as follows. The adsorption equilibrium of the organic halogen compound

142

will not be influenced to a large extent by the addition of base. On the contrary, the adsorption equilibrium of the hydrogen halide

$$H-Hal \;+\; * \;\rightleftharpoons\; \overset{\displaystyle H-Hal}{\underset{\displaystyle *}{|}}$$

has to be replaced by that for the halide anion

$$Hal^- \;+\; * \;\rightleftharpoons\; \overset{\displaystyle Hal^-}{\underset{\displaystyle *}{|}}$$

According to the hard and soft acid-base criterion[1] Hal$^-$ is much harder than HHal and will have a lower affinity for the (soft) palladium or nickel metal. In this respect, the less inhibitory effect of the hydrogen halide in polar and/or basic medium is easily understood through the higher degree of ionization.

STEREOCHEMISTRY AND MECHANISM

With palladium as the catalyst, the hydrogenolysis of the benzyl-chlorine bond occurs preferentially with retention of configuration at the benzylic carbon[2,3].

R = H, Et : > 63% inversion

Raney nickel gives almost complete racemization[3,4], *i.e.* ∿50% retention and ∿50% inversion. These results point to a chemisorption of the chlorine atom by which the carbon-halogen bond becomes situated parallel to the catalyst surface. In this manner, displacement of the chlorine by both front- and backside attack of hydrogen is possible (*cf.* the benzyl-nitrogen hydrogenolysis reaction on p 133).

Indications of a nucleophilic displacement of the halide have been inferred from the gas-phase hydrogenolysis of chlorobenzenes over palladium[5,6] and platinum[6].

In the liquid phase, however, the displacement of the halogen over both palladium and Raney nickel appears to occur by a hydrogen atom rather than a hydride[7,8] since no electronic substituent effect on the rate of hydrogenolysis has been observed.

Surprisingly, the stereochemical course of the hydrogenolysis of *gem*-dichloro-

cyclopropane derivatives (1) over nickel has been found to be strongly dependent on the nature of the base employed[9].

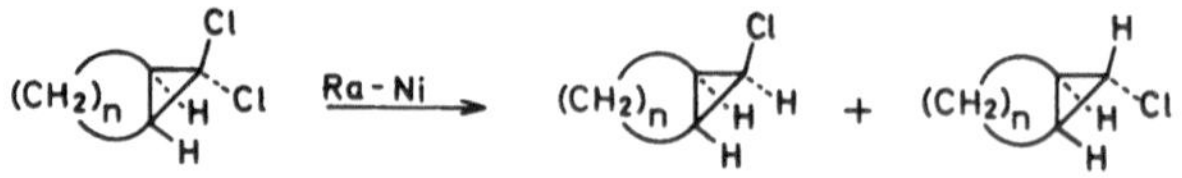

	main product using	main product using
n = 3,4,6	EtNH$_2$	
	Et$_2$NH	H$_2$N-(CH$_2$)$_n$-NH$_2$
	Et$_3$N	
	no base	(n = 2,3 or 6)

Inspection of molecular models reveals

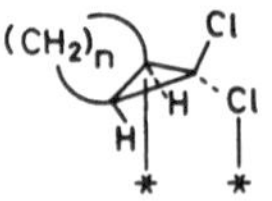

as the most favourable adsorbed state because of the contribution of both the
cyclopropane carbon-carbon bond and the chlorine for chemisorption of the
molecule. S$_N$i-type displacement of the chlorine by hydrogen from the catalyst
gives the all-*cis*-cyclopropane product when mono-amines are used as the base.
The diamines are more strongly adsorbed than the mono-amines[10] and the
above-mentioned adsorbed state may be replaced by a one-point adsorption:

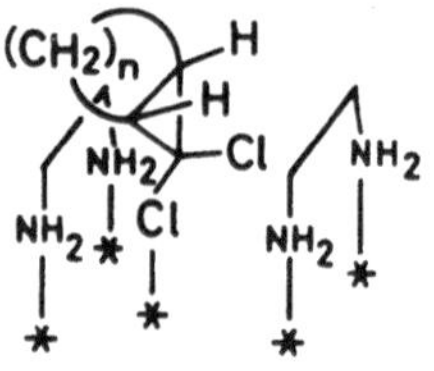

from which S$_N$i-type hydrogen attack leads to the formation of the *trans*-
cyclopropane derivative.

SELECTIVITY

Benzyl and allyl halides are hydrogenolyzed without affecting most other
reducible groups. Hydrogenolysis of vinyl, aryl and alkyl halides results in most
instances in simultaneous reduction of other functional groups, except ketonic,
aryl, and cyclopropyl groups.

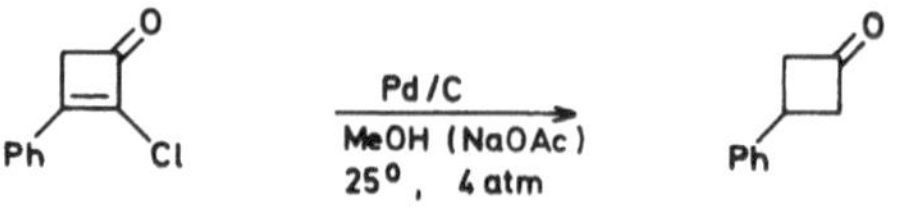

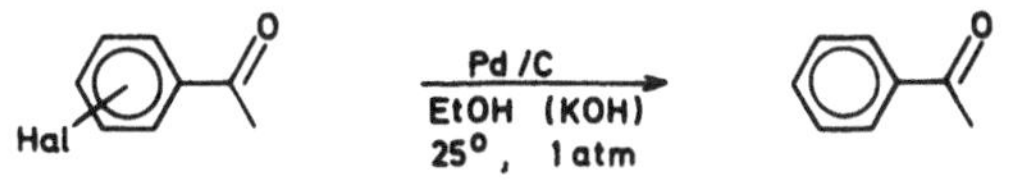

12

(Hal = I, Br, Cl; for Hal = F hydrogenolysis of the carbon-fluorine bond and hydrogenation of the ketone take place simultaneously).

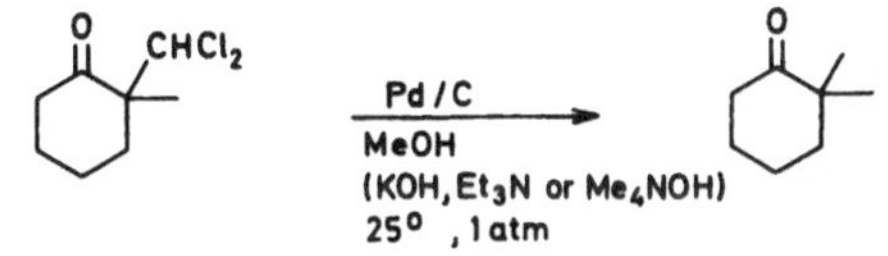

13

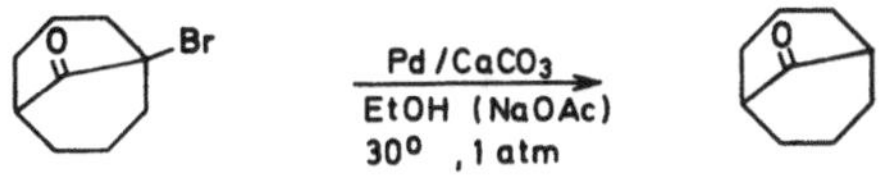

14

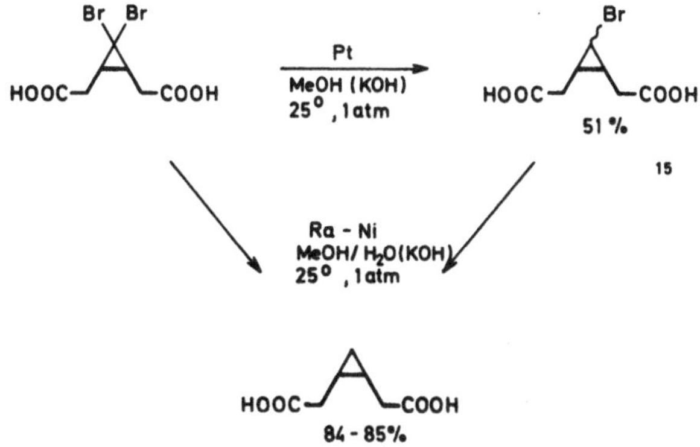

15

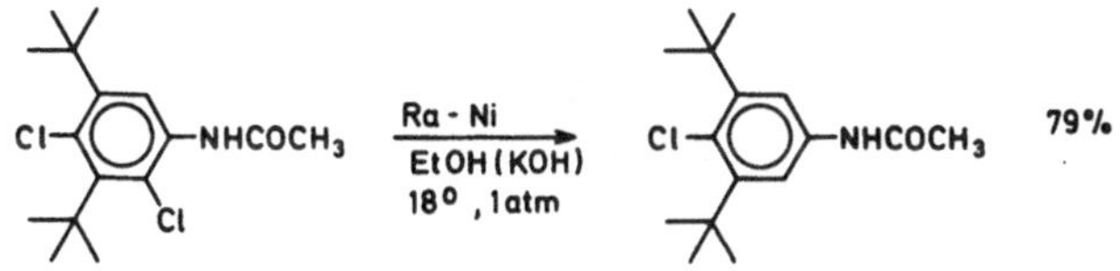

16

APPLICATION

The four examples given above show the importance of the reaction for the removal of halogens from haloketones, since other dehalogenation methods reduce the carbonyl groups as well. This selective behaviour was used for the synthesis of 5-methoxy-1-tetralone (2)[17] in which chlorine was applied as blocking group.

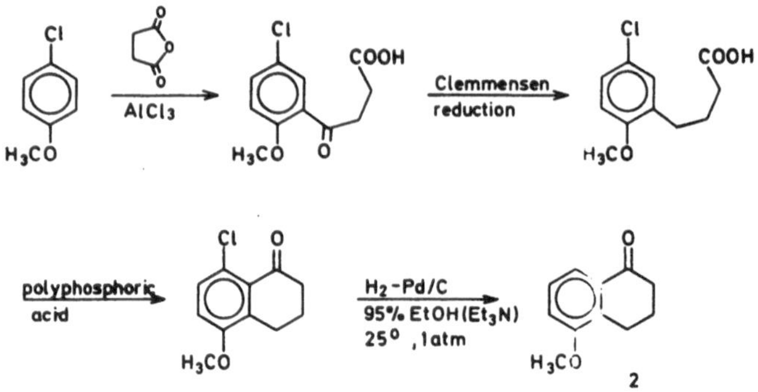

Cl / H₃CO ... (reaction scheme 1)

COOH

Clemmensen reduction

AlCl₃

polyphosphoric acid

H₂-Pd/C
95% EtOH(Et₃N)
25°, 1atm

2

Deuterolysis of the carbon-halogen bond is a convenient procedure for the synthesis of site-specific deuterium compounds, *e.g.*[18]

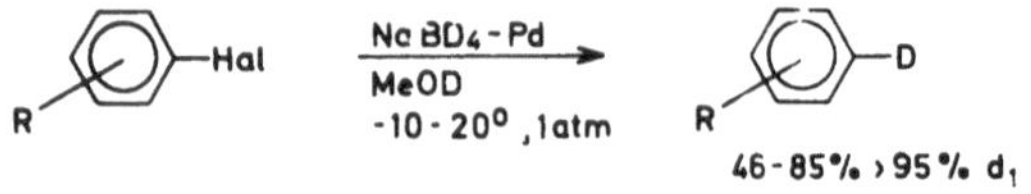

R—⟨ ⟩—Hal $\xrightarrow[\text{MeOD} \\ -10-20°, 1atm]{\text{Na BD}_4-\text{Pd}}$ R—⟨ ⟩—D

46-85% > 95% d_1

where the catalyst was prepared *in situ* by sodium borodeuteride reduction of palladium dichloride in order to exclude isotopic dilution.

Perdeuterocyclohexanols (>90% isotopic purity) were conveniently prepared by deuterolysis/deuteration of polychlorophenols over Raney nickel in deuterium oxide in the presence of NaOD at 150° and 40 atm[19].

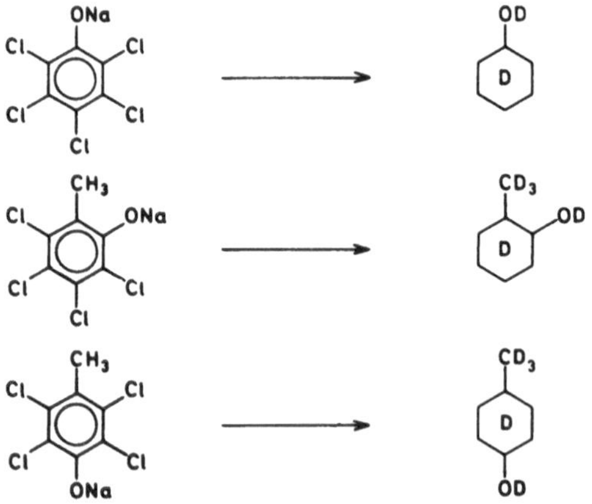

ONa / Cl,Cl,Cl,Cl,Cl → OD / D

CH₃ / Cl, ONa, Cl, Cl, Cl → CD₃ / OD, D

CH₃ / Cl, Cl, Cl, Cl, ONa → CD₃ / D, OD

Here, carbon-hydrogen deuterolysis (p 94) had also taken place. Any isotopic dilution in the product by chemisorbed hydrogen or by the solvent had been excluded, since the Raney nickel catalyst had been prepared *in situ* from nickel-aluminium alloy and deuterium oxide. The deuterium gas liberated during the catalyst preparation further served as the deuterium source (in addition to deuterium oxide) for the deuterolysis/deuteration.

146

Hydrogenolysis of the *carbonyl-chlorine bond* (the so-called *Rosenmund reduction*[20]) has found wide application in preparative organic chemistry[21] for the conversion of carboxylic acids to the respective aldehydes.

$$R-COOH \xrightarrow{SOCl_2} R-COCl \xrightarrow{H_2/Pd} R-COH$$

Although originally the reaction was carried out under rather special conditions (*i.e.* passing hydrogen gas through a solution of the acid chloride in boiling xylene with palladium on barium sulfate as the catalyst), the hydrogenolysis reaction proceeds rapidly at ambient temperature using tertiary amines or sodium acetate as acid acceptors.

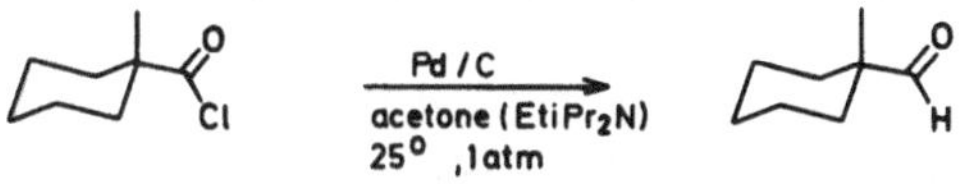

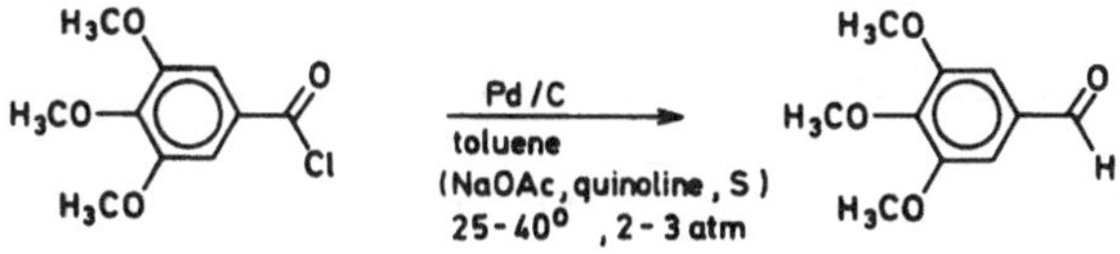

Furthermore, in the case of crowded carbonyl chlorides (3)[22], only small amounts (< 5%) of decarbonylation products were found.

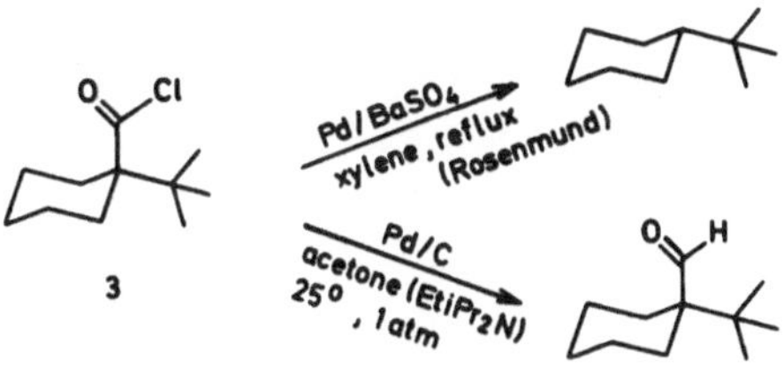

References

1. R.G. Pearson and J. Songstad, *J. Am. Chem. Soc.*, **89**, 1827 (1967).
2. E. Ott and K. Krämer, *Ber. Dtsch. Chem. Ges. B*, **68**, 1655 (1935).
3. S. Imaizumi, *Nippon Kagaku Zasshi*, **82**, 245 (1961), *Chem. Abstr.*, **57**, 9730e (1962).
4. S. Imaizumi, *Nippon Kagaku Zasshi*, **77**, 1511 (1956), *Chem. Abstr.*, **53**, 5179g (1959).
5. M. Kraus and V. Bazant, *Proc. Int. Congr. Catal. 5th, 1972*, **2**, 1073 (1973).
6. O. Hinterhofer, *Monatsh. Chem.*, **105**, 279 (1974).
7. V. Ruzicka and J. Prochazka, *Collect. Czech. Chem. Commun.*, **35**, 430 (1970).

8. C.W. Ruskamp, A.P.G. Kieboom, and H. van Bekkum, unpublished results.

9. K. Isogai, S. Kondo, K. Katsura, S. Sato, N. Yoshihara, Y. Kawamura, and T. Kazama, *Nippon Kagaku Zasshi,* **91**, 561 (1970), *Chem. Abstr.,* **74**, 3186d (1971).

10. *Cf.* A.W. van Muijlwijk, A.P.G. Kieboom, and H. van Bekkum, *Rec. Trav. Chim. Pays-Bas,* **93**, 204 (1974).

11. S.L. Manatt, M. Vogel, D. Knutson, and J.D. Roberts, *J. Am. Chem. Soc.,* **86**, 2645 (1964).

12. A.P.G. Kieboom, unpublished results.

13. M.G. Reinecke, *J. Org. Chem.,* **29**, 299 (1964).

14. A.C. Cope and D.M. Gale, *J. Am. Chem. Soc.,* **85**, 3743 (1963).

15. K. Hofmann, S.F. Orochena, S.M. Sax, and G.A. Jeffrey, *J. Am. Chem. Soc.,* **81**, 992 (1959).

16. A.J. de Koning, *Org. Prep. Proced. Int.,* **7**, 31 (1975).

17. J.W. Huffman, *J. Org. Chem.,* **24**, 1759 (1959).

18. T.R. Bosin, M.G. Raymond, and A.R. Buckpitt, *Tetrahedron Lett.,* **1973**, 4699.

19. J.D. Remijnse, H. van Bekkum, and B.M. Wepster, *Rec. Trav. Chim. Pays-Bas,* **89**, 658 (1970).

20. K.W. Rosenmund, *Ber. Dtsch. Chem. Ges.,* **51**, 585 (1918).

21. E. Mosettig and R. Mozingo, *Org. React.,* **4**, 362 (1948).

22. J.A. Peters and H. van Bekkum, *Rec. Trav. Chim. Pays-Bas,* **90**, 1323 (1971).

23. A.I. Rachlin, H. Gurien, and D.P. Wagner, *Org. Synth.,* **51**, 8 (1971).

8. Hydrogenolysis of non-carbon bonds

NITROGEN-OXYGEN BOND

Hydrogenolysis of the nitrogen-oxygen bond occurs readily with palladium, Raney nickel, platinum or rhodium as catalysts. The reaction has been most often applied to the reduction of nitro compounds to the corresponding amines. Aryl nitro derivatives are hydrogenolyzed more rapidly than nitro-alkanes and usually afford the amine in higher yield.

Mechanism

It is assumed that the reduction of the nitro group takes place in a step-wise manner.

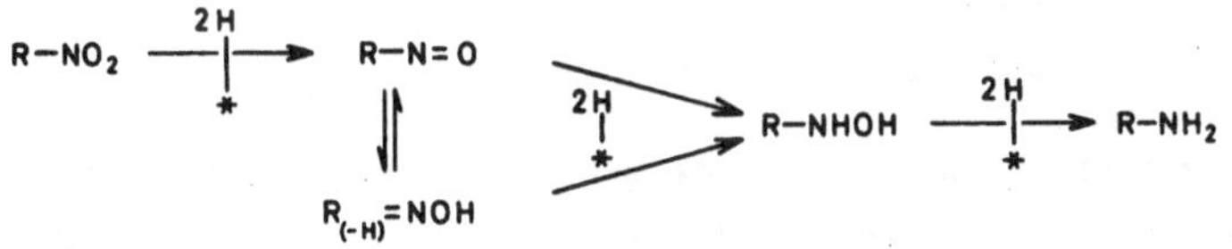

The transformation of the electron-attracting nitro to the electron-donating amino group in this reaction sequence is reflected by the positive Hammett ρ-values for the hydrogenolysis of ring-substituted nitrobenzenes[1,2] Thus *ortho*- and *para*-substituted hydroxy- and aminonitrobenzenes are hydrogenolyzed less rapidly than the corresponding *meta* derivatives because of loss of the conjugative *ortho*- and *para*-interaction in the initially adsorbed state.

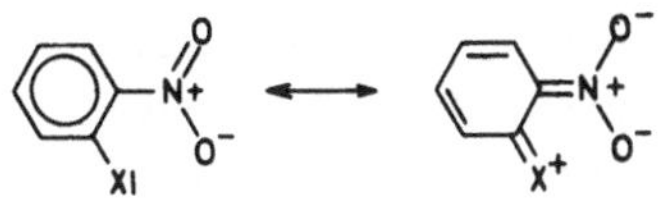

In some instances, the intermediate reduction products can be isolated, in particular for aliphatic nitro or nitroso compounds[3]. A remarkable example of this kind is[4]:

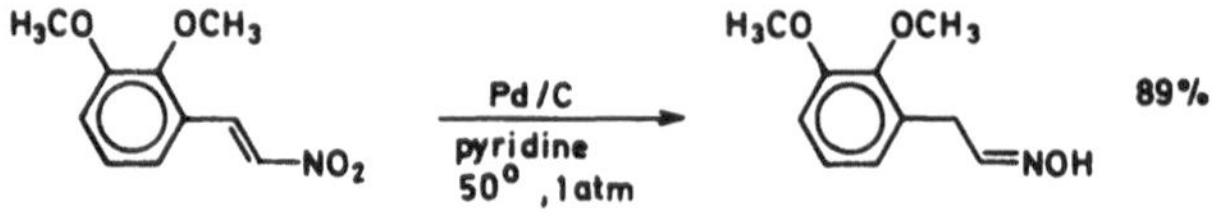

The very selective formation of the oxime after uptake of 2 equivalents of hydrogen is largely due to the very strong adsorption of the conjugated starting material as compared with the unconjugated oxime:

since further uptake of hydrogen occurs at a comparable rate yielding the corresponding amine. Similar differences in strength of adsorption between conjugated and non-conjugated compounds have been found for the hydrogenation of the ketonic carbonyl group[5]. Another example of the partial reduction of a nitro group is given by the homogeneously catalyzed hydrogenolysis of aliphatic nitro compounds[6].

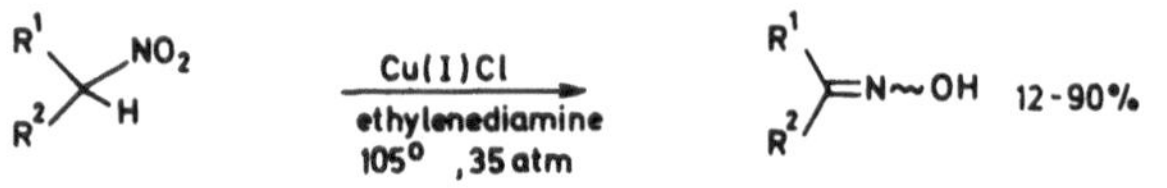

Here, the selectivity of the reaction is caused by the large difference in reaction rate between the nitro derivative and the oxime. The reaction is thought to occur via the prior formation of the nitroalkane anion.

Reactivity

As might be expected from the consecutive reaction pathway depicted above for the conversion of the nitro to the amino group, the hydrogenolysis of the nitrogen-oxygen bond of nitro, nitroso (or its tautomeric oxime), and hydroxylamino derivatives requires comparable reaction conditions.

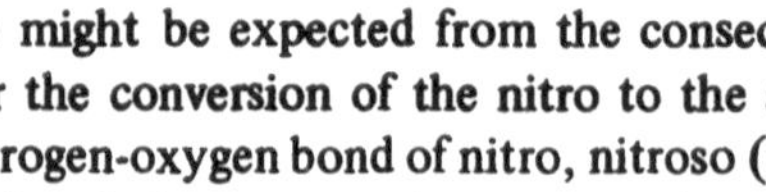

150

Ar–NO $\xrightarrow[\substack{EtOH \\ 25^\circ \text{ ,1atm}}]{Pd/C}$ Ar–NH$_2$ 7

[indane]=NOH $\xrightarrow[\substack{AcOH\ (H_2SO_4) \\ 25^\circ \text{ ,3 atm}}]{Pd/C}$ [indane]–NH$_3^+$ 8

Ph–N(OH)(R) $\xrightarrow[\substack{EtOH \\ 25^\circ \text{ ,1atm}}]{Pd/C}$ Ph–N(H)(R)   R = C$_2$H$_5$, C$_4$H$_9$ 9

The same holds for the hydrogenolysis of amine oxides, e.g.[10]:

R–[pyridine N$^+$–O$^-$] $\xrightarrow[\substack{EtOH \\ 25^\circ \text{ ,1atm}}]{Pd/C}$ R–[pyridine N]

Side reactions

Depending on the reaction conditions applied, several condensation reactions may take place during the hydrogenolysis of the nitrogen-oxygen bond[11].

$$R-N=O \ +\ R-NHOH \xrightarrow{-H_2O} R-\overset{\overset{O^-}{|}}{N^+}=N-R \xrightarrow{H_2/cat} R-\overset{H}{N}-\overset{H}{N}-R \xrightarrow[slow]{H_2/cat}$$

$$R-NH_2 \ +\ H_2N-R$$

$$\overset{R^2}{\underset{R^1}{>}}\overset{H}{\underset{}{C}}-N=O \ \rightleftharpoons\ \overset{R^2}{\underset{R^1}{>}}C=N-OH \xrightarrow{H_2NCHR^1R^2} \ \overset{R^2}{\underset{R^1}{>}}C\overset{\overset{N-OH}{|}}{\underset{\underset{R^1}{N-CH}}{<}}\overset{R^2}{\underset{H}{}}$$

$$\Big\downarrow H_2 \ cat$$

$$\overset{R^2}{\underset{R^1}{>}}\overset{H}{\underset{}{C}}-\overset{}{\underset{H}{N}}-C\overset{R^2}{\underset{R^1}{<}}H$$

This latter condensation reaction may become important in the hydrogenolysis of oximes[12], since high concentrations of both oxime and amine occur during the reduction. Hydrogenolysis of oximes is, therefore, best carried out in acid medium in order to inhibit the condensation reaction by protonation of the amine formed, whereas nitro derivatives are preferentially hydrogenolyzed in neutral (alcoholic) or slightly basic medium since the first-mentioned condensation reaction is accelerated by acids.

Selectivity

The strong adsorption of, in particular, conjugated nitro compounds and oximes in addition to both the rapid cleavage of the nitrogen-oxygen bond and a high rate of desorption of the reduction products allows, in most cases, hydrogenolysis of the nitrogen-oxygen bond to occur without affecting other reducible groups, e.g. vinyl, carbonyl, and carbon-oxygen, -nitrogen and -halogen bonds. One might say that apparently each reduced molecule is pushed away very rapidly from the catalyst surface by the unreduced nitro-containing molecules still present, i.e. not enough time is available to hydro-genate/hydrogenolyse other functional groups concomitantly. The following examples demonstrate this selectivity.

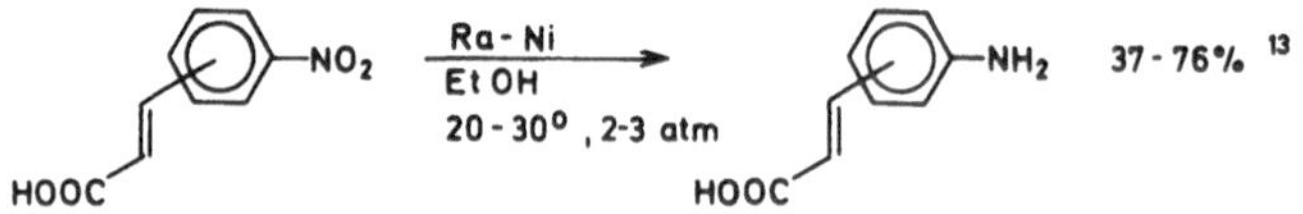

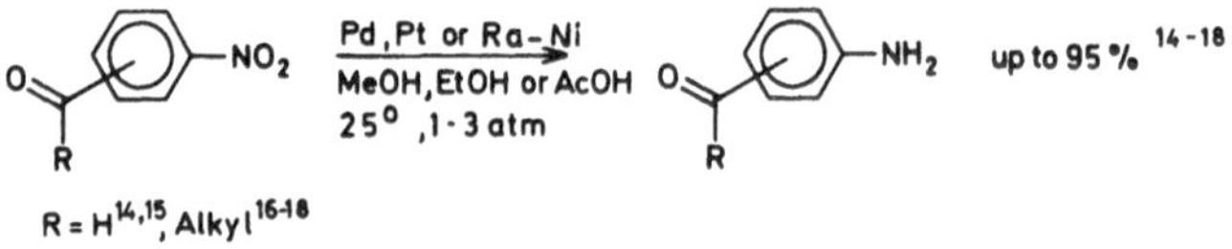

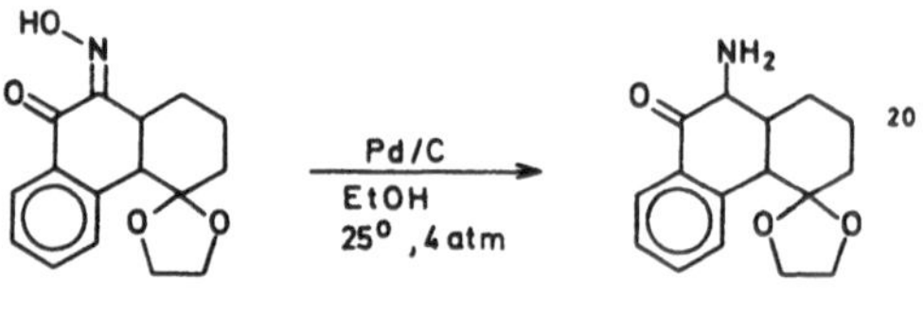

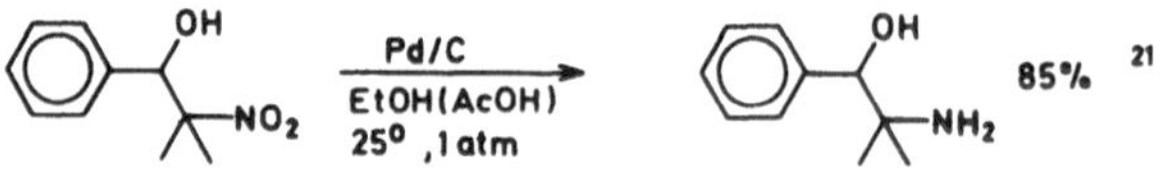

HOOC—⬡—CH=CH—⬡—NO₂ →[Ra-Ni / Et OH / 20-30°, 2-3 atm] HOOC—⬡—CH=CH—⬡—NH₂   37-76% [13]

O=C(R)—⬡—NO₂ →[Pd, Pt or Ra-Ni / MeOH, EtOH or AcOH / 25°, 1-3 atm] O=C(R)—⬡—NH₂   up to 95% [14-18]

R = H[14,15], Alkyl[16-18]

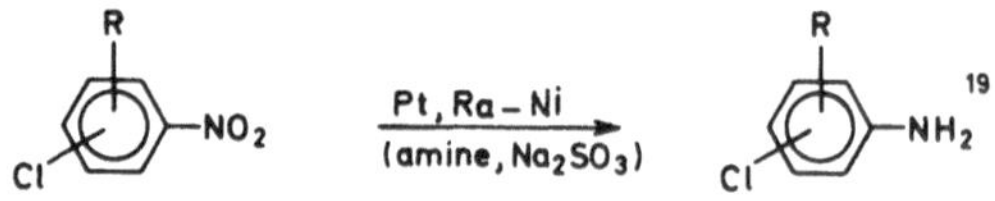

Cl—⬡(R)—NO₂ →[Pt, Ra-Ni / (amine, Na₂SO₃)] Cl—⬡(R)—NH₂   [19]

(HO-N, tetracyclic ketone/dioxolane) →[Pd/C / EtOH / 25°, 4 atm] (NH₂ substituted)   [20]

⬡—CH(OH)—CH(NO₂)(CH₃) →[Pd/C / EtOH(AcOH) / 25°, 1 atm] ⬡—CH(OH)—CH(NH₂)(CH₃)   85% [21]

The corresponding chloro derivative of the latter showed preferential carbon-chlorine hydrogenolysis[21]:

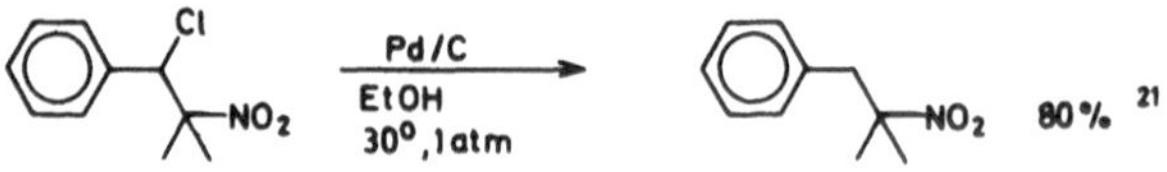

⬡—CH(Cl)—CH(NO₂)(CH₃) →[Pd/C / EtOH / 30°, 1 atm] ⬡—CH₂—CH(NO₂)(CH₃)   80% [21]

152

due to the very reactive benzyl-chlorine bond (*cf.* p 141) in addition to the sterically deactivated aliphatic nitro group. In general, aliphatic nitro compounds with other reducible functions show a smaller selectivity, due to the weaker adsorption of the nitro-part of the molecule as compared with conjugated nitro compounds. For instance, nitroalkenes may be converted to the corresponding nitroalkanes. Selective conversion of dinitro into aminonitro derivatives may occur by interrupting the reaction after the uptake of three equivalents of hydrogen. The formation of the aminonitro compound as the main product is mainly due to the much stronger adsorption of the dinitro with respect to the aminonitro compound. Which nitro function will be hydrogenolyzed first depends on both steric and electronic effects.

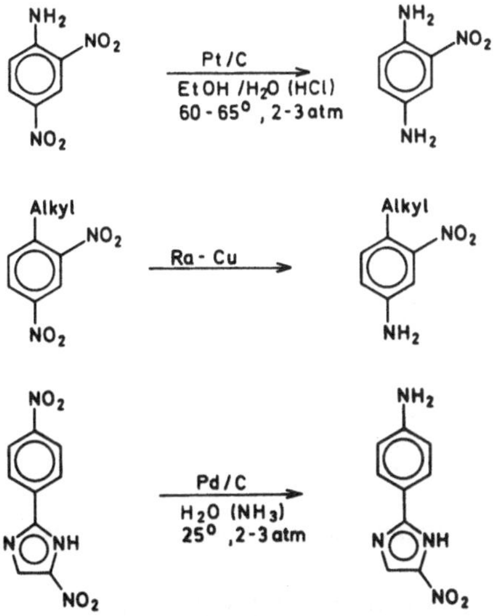

As expected, the least hindered nitro group is hydrogenolyzed preferentially, whereas electron donation by the adjacent group retards the reaction (see p 149). Furthermore, aryl nitro compounds are adsorbed more strongly as well as hydrogenolyzed more readily than aliphatic nitro compounds, which in most cases allows selective reduction of the aryl nitro group.

$$O_2N \sim\!\!\sim\!\!\sim Ar\!-\!NO_2 \longrightarrow O_2N \sim\!\!\sim\!\!\sim Ar\!-\!NH_2$$

Application

In conclusion, the heterogeneously catalyzed hydrogenolysis of the nitro group is often the method of choice as a reduction technique, in particular for reactants containing other reducible functional groups. It may be noted that simultaneous reduction of a nitro and a carbonyl group may result in considerable amounts of undesired by-products[17]. For instance, the direct conversion of *m*-nitrobenzoyl derivatives (1) according to[26,27]

1

requires an acidic medium in order to achieve the carbon-oxygen hydrogeno-lysis. The presence of acid promotes both the formation of hydrazines (p 151) and the reductive alkylation of the amine group (p 116) during the hydrogenolysis reaction. Therefore, much better results are obtained using the following – longer – reaction sequence[17]:

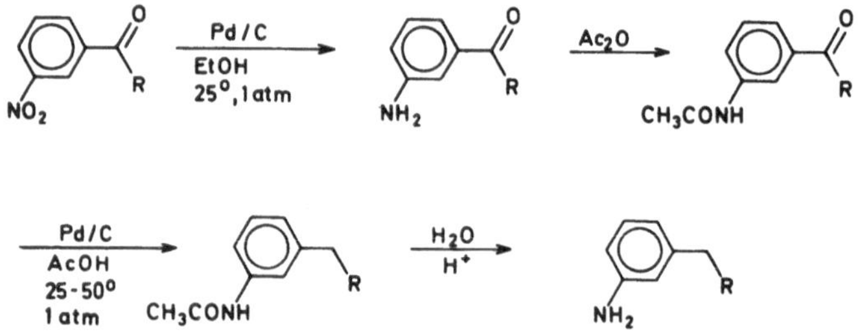

Acylation of the amine prevents the undesired reductive alkylation reaction. On the other hand, the facile one-step indole (2) synthesis according to[28]

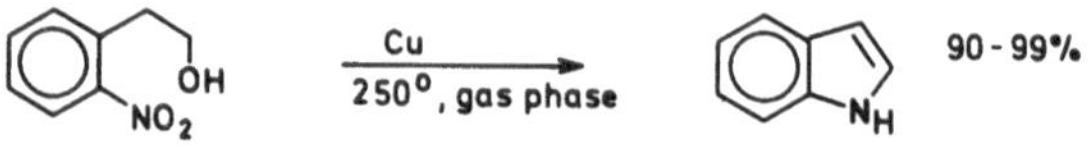

is an example of a desired reductive alkylation reaction during the reduction of the nitro group.

Recently, the homogeneous dichlorotris(triphenylphosphine)ruthenium(II) catalyst has been found to hydrogenolyse dinitroaromatics to the corresponding aminonitro compounds in good yields[25].

NITROGEN-NITROGEN BOND

Hydrogenolysis of the N-N bond occurs under mild conditions using palladium, platinum or Raney nickel as the catalyst. In this way azides (3), hydrazones (4), hydrazines (5), azo compounds (6) etc. are conveniently converted to the respective amines.

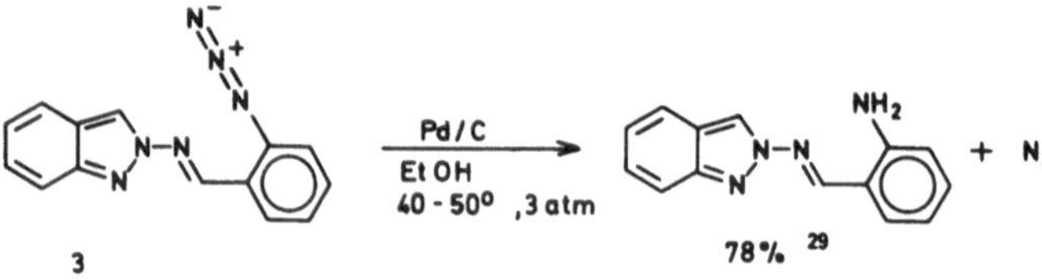

3

78% [29]

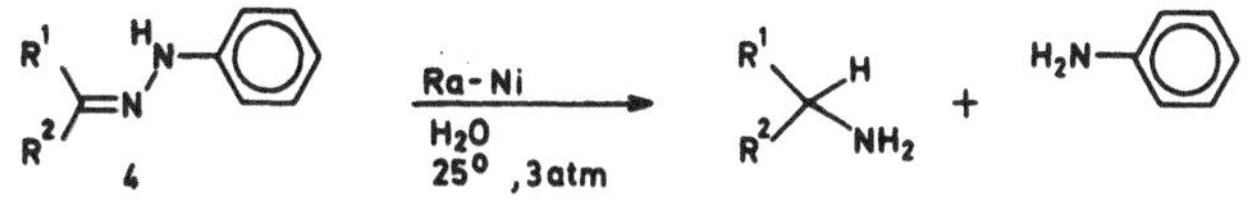

(in the synthesis of amino compounds from monosaccharides)

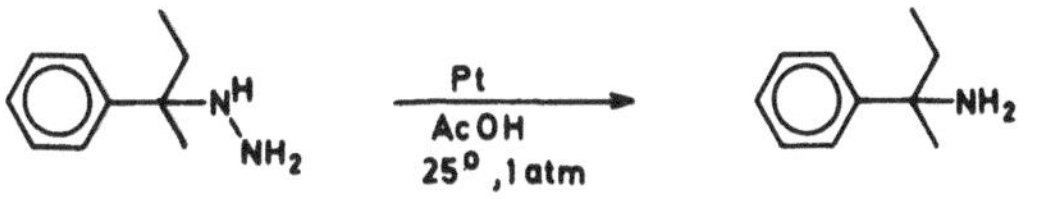

OXYGEN-OXYGEN BOND

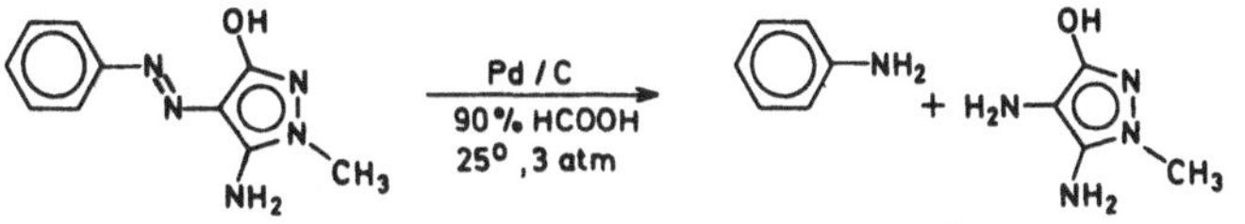

Hydrogenolysis of the O-O bond occurs very rapidly in (hydro)peroxides and ozonides.

$$R^1-O-O-R^2 \longrightarrow R^1-OH \ + \ HO-R^2$$

The hydrogenolysis of (hydro)peroxides over palladium and platinum affords the alcohols (and water) in excellent yields without, in most instances, affecting other reducible groups, whereas the reduction of ozonides may give rather complex reaction mixtures[33]. The hydrogenolysis of ozonides in basic medium, however, shows a much higher selectivity towards aldehyde formation, e.g.[34]

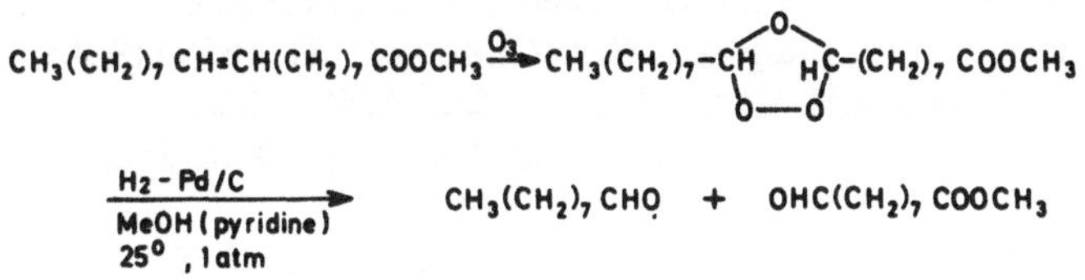

The base inhibits any acid-catalyzed decomposition of the ozonide and retards further hydrogenation of the aldehydes formed.

SULPHUR-OXYGEN BOND

Reduction of sulphoxides with hydrogen over palladium affords the corresponding sulphides in good yield leaving carbon-oxygen and, to some extent, carbon-carbon double bonds intact[35].

155

$$R^1-\overset{\overset{\text{O}}{\|}}{S}-R^2 \quad \xrightarrow[\substack{\text{EtOH, 80-90}^\circ \\ \text{30 - 90 atm}}]{\text{Pd/C}} \quad R^1-S-R^2$$

$R^1 = R^2 = C_4H_9$, $C_6H_5CH_2$, C_6H_5 (90–99%)

$R^1, R^2 = (CH_2)_2 \overset{\overset{\text{O}}{\|}}{C}(CH_2)_2$ (59%)

$R^1 = 4\text{-}CH_3C_6H_4$, $R^2 = C_6H_5CH=CH$ (66%)

In the latter case, the use of an excess amount of catalyst results in additional hydrogenation of the carbon-carbon double bond.

References

1. A.V. Finkelshtein, *Reakts. Sposobn. Org. Soedin.*, 3, 47, 72, 126 (1966); A.V. Finkelshtein and Z.M. Kuzmina, *Dokl. Akad. Nauk USSR*, 158, 176 (1964) and 171, 915 (1966), *Chem. Abstr.*, 62, 3900a (1965) and 66, 64964n (1967).
2. V. Ruzicka and H. Santrochova, *Collect. Czech. Chem. Commun.*, 34, 2999 (1969).
3. *Cf.* J. Meinwald, Y.C. Meinwald, and T.N. Baker, *J. Am. Chem. Soc.*, 86, 4074 (1964).
4. A. Lindenmann, *Helv. Chim. Acta*, 32, 69 (1949).
5. H. van Bekkum, A.P.G. Kieboom, and K.J.G. van de Putte, *Rec. Trav. Chim. Pays-Bas*, 88, 52 (1969).
6. J.F. Knifton, *J. Catal.*, 33, 289 (1974).
7. W.T. Sumerford, W.H. Hartung, and G.L. Jenkins, *J. Am. Chem. Soc.*, 62, 2082 (1940).
8. W.E. Rosen and M.J. Green, *J. Org. Chem.*, 28, 2797 (1963).
9. G.E. Utzinger and F.A. Regenass, *Helv. Chim. Acta*, 37, 1885 (1954).
10. A.R. Katritsky and A.M. Monro, *J. Chem. Soc.*, 1958, 1263.
11. D.V. Sokolskii and V.P. Chmonina, *Proc. Int. Congr. Catal. 2nd, 1960*, 2, 2733 (1961).
12. P.N. Rylander and D.R. Steele, *Engelhard Ind. Tech. Bull.*, 5, 113 (1965), *Chem. Abstr.*, 63, 5527d (1965).
13. E.R. Blout and D.C. Silverman, *J. Am. Chem. Soc.*, 66, 1442 (1944).
14. W. Borsche and W. Ried, *Ber. Dtsch. Chem. Ges. B*, 76, 1011 (1943); W. Borsche and F. Sell, *Chem. Ber.*, 83, 78 (1950).
15. W. Ried, A. Berg, and G. Schmidt, *Chem. Ber.*, 85, 204 (1952).
16. N.J. Leonard and S.N. Boyd, Jr., *J. Org. Chem.*, 11, 405 (1946).
17. N.L. Allinger and E.S. Jones, *J. Org. Chem.*, 27, 70 (1962); A.P.G. Kieboom, unpublished results.
18. H.M.A. Buurmans, B. van de Graaf, and A.P.G. Kieboom, *Org. Mass Spectrom.*, 5, 1081 (1971).

19. K. Habig and K. Baessler, German Patent, 2,105,682 (1972), *Chem. Abstr.*, 77, 151651d (1972); Y. Hirai and K. Miyata, German Patent, 2,308,105 (1972), *Chem. Abstr.*, 79, 126058z (1973); Y. Hirai and K. Miyata, Japan. Patent, 73 49,728 (1973), *Chem. Abstr.*, 79, 136770q (1973).

20. D. Ginsburg and R. Pappo, *J. Chem. Soc.*, 1953, 1524.

21. F.H. Marquarat, *Ann. N.Y. Acad. Sci.*, 214, 110 (1973).

22. W.H. Brunner and A. Halasz, U.S. Patent, 3,088,978 (1963), *Chem. Abstr.*, 60, 2826d (1964).

23. W.H. Jones, W.F. Benning, P. Davis, D.M. Mulvey, P.I. Pollak, J.C. Schaeffer, R. Tull, and L.M. Weinstock, *Ann. N.Y. Acad. Sci.*, 158, 471 (1969).

24. W.H. Jones, S.H. Pines, M. Sletzinger, *Ann. N.Y. Acad. Sci.*, 214, 150 (1973).

25. J.F. Knifton, *J. Org. Chem.*, 41, 1200 (1976).

26. H. Oelschläger, *Chem. Ber.*, 89, 2025 (1956).

27. I.D. Pletneva, R.S. Muromova, and I.V. Shkhiyants, *Zh. Vses. Khim. Ova.*, 10, 595 (1965); *Chem. Abstr.*, 64, 1975d (1966).

28. J. Bakke, H. Heikman, and E.B. Hellgren, *Acta Chem. Scand., Ser. B*, 28, 393 (1974).

29. L. Krbechek and H. Takimoto, *J. Org. Chem.*, 29, 1150 (1964).

30. M.L. Wolfrom, F. Shafizadeh, J.O. Wehrmüller, and R.K. Armstrong, *J. Org. Chem.*, 23, 571 (1958).

31. D.J. Cram and J.S. Bradshaw, *J. Am. Chem. Soc.*, 85, 1108 (1963).

32. E.C. Taylor, J.W. Barton, and T.S. Osdene, *J. Am. Chem. Soc.*, 80, 421 (1958).

33. P.S. Bailey, *Chem. Rev.*, 58, 925 (1958).

34. E.H. Pryde, D.E. Anders, H.M. Teeter, and J.C. Cowan, *J. Org. Chem.*, 27, 3055 (1962).

35. K. Ogura, M. Yamashita, and G. Tsuchihashi, *Synthesis*, 1975, 385.

The book gives a comprehensive description of the scope of
catalytic hydrogenation and hydrogenolysis as synthetic
organic methods. After a short introduction concerning the
catalyst system and reaction conditions, hydrogenation and
hydrogenolysis are treated separately using numerous
synthetically interesting examples. The selectivity and
stereochemistry of the hydrogen addition to the various bonds
has been dealt with in relation to the reaction mechanism. In a
similar way, the scope and limitations of the reactions have
been discussed. The book aims at providing preparative
organic chemists with the insight and know-how necessary to
apply catalytic hydrogenation and hydrogenolysis to synthetic
problems.

A.P.G. Kieboom and F. van Rantwijk are
staff-members of the Laboratory of Organic Chemistry,
Department of Chemical Technology, Delft University of
Technology, Delft, The Netherlands.

Delft University Press